Municipal Waste Incineration Risk Assessment

Deposition, Food Chain Impacts, Uncertainty, and Research Needs

CONTEMPORARY ISSUES IN RISK ANALYSIS

Sponsored by the Society for Risk Analysis

Municipal Waste Incineration Risk Assessment

Deposition, Food Chain Impacts, Uncertainty, and Research Needs

Edited by
Curtis C. Travis

Oak Ridge National Laboratory
Oak Ridge, Tennessee

Plenum Press • New York and London

Library of Congress Cataloging in Publication Data

United States Environmental Protection Agency/Oak Ridge National Laboratory
Workshop on Risk Assessment for Municipal Waste Combustion: Deposition, Food
Chain Impacts, Uncertainty, and Research Needs (1989: Cincinnati, Ohio)
    Municipal waste incineration risk assessment: deposition, food chain impacts, uncer-
tainty, and research needs / edited by Curtis C. Travis.
        p.   cm. — (Contemporary issues in risk analysis; v. 5)
    "Proceedings of a United States Environmental Protection Agency/Oak Ridge Na-
tional Laboratory Workshop on Risk Assessment for Municipal Waste Combustion:
Deposition, Food Chain Impacts, Uncertainty, and Research Needs, held June 8–9,
1989, in Cincinnati, Ohio" — T.p. verso.
    Includes bibliographical references and index.
    ISBN 0-306-44016-4
    1.   Incineration — Health aspects — Congresses. 2.   Incineration — Environmental
aspects — Congresses.   3. Health risk assessment — Congresses. I. Travis, C. C. II. United
States. Environmental Protection Agency. III. Oak Ridge National Laboratory. IV.
Title. V. Series.
    [DNLM: 1. Air Pollution — prevention & control — United States — congresses. 2.
Refuse Disposal — methods — congresses. 3. Risk Factors — congresses. 4. Waste Products —
adverse effects — congresses. WA 780 U567m 1989]
RA578.H38U55   1989
628.5'3 — dc20
DNLM/DLC   91-24094
for Library of Congress   CIP

Proceedings of a United States Environmental Protection Agency/
Oak Ridge National Laboratory Workshop on Risk Assessment for
Municipal Waste Combustion: Deposition, Food Chain Impacts,
Uncertainty, and Research Needs, held June 8–9, 1989,
in Cincinnati, Ohio

ISBN 0-306-44016-4

© 1991 Plenum Press, New York
A Division of Plenum Publishing Corporation
233 Spring Street, New York, N.Y. 10013

PREFACE

The disposal of large quantities of municipal solid waste (MSW) being generated by industrialized countries has become a serious problem. Since it is estimated that within 10 years, half of all municipalities will lack sufficient landfill space, many cities are considering municipal waste combustion as an alternative waste management option.[1] Municipal waste combustors have been a source of contention in many local communities and a growing research topic in the scientific community. This book represents a compilation of chapters written by experienced individuals in the areas of emissions estimation, deposition modeling, risk assessment, indirect exposures, and uncertainty analysis.

Estimation of potential human risks associated with pollutants has become an increasing concern. Most often, values required for deposition rates and annual atmospheric concentrations are estimated through the use of atmospheric dispersion models. Chapter 1 compares data on the flat terrain versus the complex terrain dispersion models such as the U.S. EPA Industrial Source Complex Short Term (ISCST) and Long Term (ISCLT).

Chapter 2 focuses on the modeling of atmospheric dispersion and dry deposition of fine particulates. A specific size particle (10-20 um) is used because of its relevance to municipal waste facilities since best available control technology effectively removes particulates above this size range.

The deposition of materials from the atmosphere is a critical link in the pathway by which toxic atmospheric pollutants are transported to the surface of food chain components. Chapter 3 describes the importance accounting for wet deposition in risk assessments of municipal waste incinerators.

v

Accounting for dry deposition in incinerator risk assessments is also critical. Travis and Hattemer-Frey (1987) found this particular pathway to be the major source of human exposure to background levels of dioxin. Chapter 4 evaluates the effectiveness of three different particle size-dependent deposition models and the classical particle size-independent model.

Corresponding with the previous two chapters, Chapter 5 involves the long-range transport and deposition of "semivolatile" organic compounds (SOCs), including the distribution of the SOCs between the gas and particle phases in ambient air, and the relationship of this distribution to the wet removal of SOCs from the atmosphere.

Among the health impacts from municipal waste incinerators, of particular concern are potential exposures from ingesting contaminated food items. Chapter 6 quantifies the extent of human exposure to 2,3,7,8-tetrachlorinated dibenzo-p-dioxin (TCDD) and cadmium emitted from a typical MWC in the U.S. It also provides an innovative perspective on human exposure to facility-emitted pollutants using a probabilistic risk assessment approach.

One of the first steps in the assessment of human exposure to an organic contaminant is the estimation of the chemical in various environmental media. Chapter 7 focuses on the accumulation of pollutants via the plant route using a fugacity-based analysis and a kinetic model of uptake and distribution of organic compounds in plants. Chapter 8 continues the discussion of the plant route by considering the air-to-leaf transfer of organic vapors to plants.

The fate of a chemical in the plant environment depends on both the properties of the chemical as well as the nature of the plant or plant community. Chapter 9 discusses some unifying concepts which bring order to the understanding of the complex relationships between chemical and plant. Also, Chapter 9 includes a discussion of a model that describes the uptake, translocation, accumulation, and biodegradation of organic chemicals in terrestrial plants.

A realistic strategy for managing the potential health risks of MWC emissions requires adequate attention to uncertainties. Estimates of the level of exposure to MSW incinerator emissions can be improved by direct measurements of contaminant concentrations in various media; however, it is almost always necessary to rely on models to estimate these concentrations. Chapter 10 elucidates some of the uncertainties involved in estimating chemical degradation and accumulation in the environment. Specific examples include polychlorinated dibenzo-p-dioxins and dibenzofurans.

Chapter 11 assesses the magnitude of human exposure through the food chain for two pollutants released by MWCs: cadmium and 2,3,7,8-tetrachlorodibenzo-p-dioxin. This chapter also considers the sensitivity of model predictions to uncertainties in model input parameters and the extent to which variability in exposure estimates is attributable to uncertainty in the terrestrial food chain model.

Chapter 12 examines the completeness of the exposure model and the treatment of uncertainty in exposure estimates. Using contaminant transfers from air to food as a case study, this chapter distinguishes between variability, ignorance, and uncertainty.

Dealing specifically with waste-to-energy facilities in California, Chapter 13 discusses the uncertainties involved in quantifying health risks. Since quantitative health risk assessments have become the accepted method for evaluating the potential health risks associated with public exposure to pollutants, Chapter 13 focuses on the four components of such a risk assessment and the uncertainties particular to each one.

Because of the recent concern over levels of dioxin in the environment, Chapter 14 considers evidence that dioxin levels in the sediment levels have risen. Chapter 14 presents the results of a program to determine the sources, occurrence, and effect of dioxin concentrations in several locations in Ohio.

Chapter 15 also considers dioxin and the role of MWCs in dioxin emissions. This chapter examines various issues concerning incinerator emissions, including the role of uncertainty and background exposures in the assessment of health risks.

Connecticut is the first state in the country to have adopted an ambient air quality standard for dioxins. Chapter 16 describes the scientific basis and methodology used by the State Department of Health Services to establish a health-based dioxin standard.

It was the editor's intent that the presentation and interpretation of information presented herein be useful to the scientist, the layperson, those individuals whose incineration-related activities are regulated, and to those who make regulatory decisions. While much remains to be learned about the human health risks associated with MSW incineration, it is the editor's goal that the material presented in this book stimulate further research and encourage readers to formulate opinions about some of the fundamental issues affecting the management of municipal solid waste.[2]

Curtis C. Travis

REFERENCES

1. U.S. EPA. (1987). Municipal Waste Combustion Study: Characterization of the Municipal Waste Combustion Industry, EPA/530-SW-87-021H, Pollutant Assessment Branch, Research Triangle Park, NC.

2. Hattemer-Frey, H.A. and Curtis C. Travis, eds. (1991). *Health Effects of Municipal Waste Incineration*. Boca Raton, Florida: CRC Press.

CONTENTS

EVALUATION OF FLAT VERSUS COMPLEX TERRAIN MODELS
IN ESTIMATING POLLUTANT TRANSPORT AND DEPOSITION
IN COMPLEX TERRAIN

Mark W. Yambert, Greg D. Belcher, Curtis C. Travis

Office of Risk Analysis
Health and Safety Research Division
Oak Ridge National Laboratory, Oak Ridge, TN

INTRODUCTION

Estimation of potential human health risks associated with pollutants emitted into the atmosphere from the stacks of municipal waste combustors (MWCs), fossil fuel plants, and similar point sources requires accurate knowledge of annual average atmospheric pollutant concentrations and deposition rates in the areas surrounding these facilities. Values for these quantities are most often estimated through the use of atmospheric dispersion models. While a vast number of these models exist and are readily available for use, the risk assessor is generally faced with obstacles such as little or no data on the size and distribution of emitted particles, limited computing capability, crude meteorological measurements, or insufficient knowledge of terrain for the location being modeled. Consequently, common practice when determining exposure to pollutants emitted from these facilities has been to use air dispersion models which require minimal computing power and site specific information. The majority of these models are based on the Gaussian plume formulation and contain only limited provisions for dealing with complicated atmospheric processes such as pollutant deposition and flow in complex terrain (i.e. terrain exceeding the height of the stack). Of these, undoubtedly the most widely used are the U.S. Environmental Protection Agency (EPA) recommended Industrial Source Complex Short Term (ISCST) and Long Term (ISCLT) models (U.S. EPA, 1979).

As use of ISCST/LT has grown, concern has been expressed regarding just how well these models predict atmospheric pollutant concentrations and deposition rates. For example, flat terrain models such as ISC are often substituted for more advanced models recommended by U.S. EPA (1986a) when modeling areas of complex terrain. In addition, ISC has also been criticized for overestimating pollutant deposition rate (Tesche et al., 1987). As a means of determining the validity of these concerns, an evaluation was performed by the Office of Risk Analysis of the Oak Ridge National

Laboratory between two flat terrain models, ISCST and ISCLT, and two current, state-of-the-art complex terrain models recommended by U.S. EPA (1986a), COMPLEX I and the Rough Terrain Diffusion Model (RTDM), that were modified to estimate pollutant deposition. The two primary questions for the study were: (1) is it acceptable to use flat terrain models such as ISCST and ISCLT to model sources located in complex terrain; and (2) are more sophisticated, particle size-dependent deposition models better than algorithms currently used in ISCST and ISCLT? This report contains brief descriptions of the modifications made to COMPLEX I and RTDM, and the findings and conclusions from the study.

DESCRIPTION OF MODELS USED IN THE STUDY

ISC Models

   The short and long term versions of the Industrial Source Complex Model, ISCST, and ISCLT are preferred for use in flat or moderately rolling terrain when modeling sources such as stacks from MWCs or other industrial source complexes (U.S. EPA, 1986a). Both are steady state Gaussian plume models which contain limited provisions for estimating the effects of terrain on pollutant transport. ISCST and ISCLT are sometimes used to estimate pollutant transport in complex terrain. However, concentrations estimated for points above stack top elevation are subject to extreme uncertainty as the models substitute stack top elevation for the point's actual elevation. Pollutant deposition is calculated through the use of surface reflection coefficients and gravitational settling velocities.

COMPLEX I

   COMPLEX I is a recommended second level screening model for use in complex terrain (U.S. EPA, 1986a). Like ISCST, it is also a Gaussian plume model. For terrain below the height of the stack being modeled, the algorithms used by COMPLEX I are essentially the same as those found in ISCST. However, for terrain above stack height COMPLEX I contains five traditional options for modeling complex terrain effects (Turner, 1986). The recommended option makes use of a terrain correction factor which reduces pollutant concentrations for elevations up to 400 meters above stack height (Turner, 1986). At elevations above this height, concentrations are set equal to zero.

   Algorithms used in COMPLEX I to estimate atmospheric pollutant concentrations are not capable of calculating pollutant deposition. Since one of the chief goals of this study was to determine whether deposition rates estimated by a more sophisticated deposition model were a significant improvement over those estimated by ISCST, it was necessary to modify the concentration algorithms of COMPLEX I to calculate pollutant deposition. Since concentration and deposition are interdependent (i.e., plume depletion), accurate representations for both of these quantities can only be obtained from concentration algorithms which consider depositional effects. An evaluation of several existing

2

models capable of estimating pollutant deposition revealed that the algorithms found in the Multiple Point Source Algorithm with Terrain Adjustments Including Deposition and Sedimentation (MPTER-DS) model (Rao et al., 1982), which are formulated on the assumption that pollutant deposition can be calculated from the product of ground level concentration and a deposition velocity, fully met this criterion. For the purposes of this analysis, these algorithms were substituted for the concentration algorithms originally found in COMPLEX I. The resulting complex terrain deposition model will hereafter be referred to as COMPDEP.

The algorithms from Rao et al. (1982) estimate pollutant deposition based on a single deposition velocity that is independent of particle size or atmospheric conditions. In reality, deposition velocity is highly dependent on particle size and atmospheric conditions. As a means of improving estimates of pollutant deposition rates, deposition velocities for the COMPDEP model are automatically calculated using computerized routines developed by the California Air Resources Board (CARB, 1986) based on the work of Sehmel (1980). These algorithms were selected based on results from an evaluation of four deposition velocity models (Travis et al., 1989) which showed them to be superior in estimating deposition velocities of particulates to grasses and small crops, the type of surfaces in risk assessments for which deposition estimates are often desired.

One final feature added to the COMPDEP model is the capability of accounting for building wake effects. COMPLEX I does not contain a methodology for estimating building wake effects. Therefore, to provide COMPDEP with this capability, algorithms from the ISCST model were used.

RTDM

RTDM is a third-level screening model for use in complex terrain (U.S. EPA, 1986a). Like COMPLEX I, RTDM modifies the effective plume height to account for the effects of terrain. However, RTDM accomplishes this in a more sophisticated manner by utilizing two additional features not found in COMPLEX I. The first through the use of a critical plume height. Critical height for a terrain feature is the height above which air passes over the terrain and below which air impacts terrain (ERT, 1987).

The second additional feature used by RTDM is the modification of pollutant concentrations to account for the effects of partial plume reflection from sloping terrain. Conventional Gaussian plume models model the ground surface through the use of "image source" routines. These routines can result in unrealistically high pollutant concentration estimates for cases in which a plume approaches rapidly sloping terrain. The partial plume reflection algorithm used in RTDM prevents these inaccuracies by multiplying pollutant concentrations by a reflection factor, R, based on terrain slope and plume growth. This factor is the ratio of the minimum value of the maximum cross wind integrated concentration (MCWI) to the cross wind integrated concentration evaluated at the point of impact of closest plume approach to terrain (ERT, 1987).

Table 1. Source Characteristics for MWC Analysis

Parameter	Value
Stack diameter (m)	1.04
Stack exit velocity (m/s)	15.24
Stack gas temperature (K)	327.60
Building Dimensions (m)	
Height	11.00
Width	48.80
Length	73.20

Like COMPLEX I, it was necessary to modify RTDM to include this capability since it contains no provision for estimating pollutant deposition rate. The modifications performed in developing the new model, RTDMDEP, with the exception of the inclusion of building wake effects, are identical to those described for COMPLEX I.

ANALYSIS PROCEDURES

Model runs were made for a hypothetical complex terrain incinerator using COMPDEP, RTDMDEP, ISCST, and ISCLT. Each model was run in two modes. The first of these determined atmospheric pollutant concentrations assuming no deposition occurred, while the second calculated both atmospheric pollutant concentrations and pollutant deposition rates. Model input is described in the following sections.

Model Input

Source Characteristics: The source parameters for the hypothetical incinerator used in this study, given in Table 1., are taken from an existing mass burn incinerator located in Rutland, Vermont. Unit emission rates of pollutant were used for all model runs. For model runs which considered pollutant deposition, emitted pollutants were assumed to fall into one of three particle size categories: (1) less than 2 μm, (2) 2 to 10 μm, and (3) greater than 10 μm. Representative sizes and fraction of emissions in each of these categories were based on data for organic pollutants found in U.S. EPA (1986b) and are, respectively, 1.0 μm and 0.875 for category 1, 6.78 μm and 0.095 for category 2, and 20.0 μm and 0.030 for category 3.

Meteorological Data: In order to consider realistic meteorological conditions, model runs were made using actual meteorological data. ISCST, COMPDEP and RTDMDEP require meteorological data in standard, preprocessed format containing hourly values for mixing height, stability class, temperature, wind direction, and wind speed. Data for the year 1970 from station number 14735 in Albany, New York were provided by the Source Receptor Analysis Branch of the U.S. EPA at Research Triangle Park, NC.

Unlike the former models, ISCLT requires meteorological data in annual stability array (STAR) summary format. The data list the frequency of occurrence of each combination of Pasquill atmospheric stability category, wind speed and wind direction over a period of one or more years. STAR data for 1970, stored on magnetic tape at the Oak Ridge National Laboratory, were used for the analysis. In addition, ISCLT requires values for mixing height and temperature. Mixing height and temperature data for Albany were taken from Holzworth (1972) and Gale Research Co. (1983), respectively.

Receptor Locations: Receptor locations used by ISCST/LT and COMPDEP model runs were chosen to reflect how model predictions are influenced by terrain near to far away from the source. Whereas terrain elevation is the chief aspect of complex terrain which affects predictions made by COMPDEP and ISCST/LT, predictions made by RTDMDEP are also influenced by hill slope and height. In order to analyze differences in model predictions due to these quantities, variable terrain slopes were used. Predictions made by COMPDEP and ISCST/LT were calculated for a variety of terrain elevations, with no consideration given to terrain slope. Receptor distances for these model runs varied from 200 to 50000 meters while elevations ranging from that of the stack base to 300 meters above the of the stack were considered. The most frequent wind direction for all stability classes during 1970 was due north (U.S. EPA, 1987). Thus, to ensure exposure to the widest possible range of meteorological conditions, all receptors were located along this direction.

Table 2. Receptor Locations Used By COMPDEP and ISCST

Receptor Number	X Coordinate (m)	Y Coordinate (m)	Elevation (m)
1	0.0	200.0	169.0
2	0.0	400.0	169.0
3	0.0	500.0	169.0
4	0.0	600.0	169.0
5	0.0	800.0	169.0
6	0.0	1000.0	169.0
7	0.0	1250.0	169.0
8	0.0	1500.0	169.0
9	0.0	1750.0	169.0
10	0.0	2000.0	169.0
11	0.0	2500.0	169.0
12	0.0	3000.0	169.0
13	0.0	4000.0	169.0
14	0.0	5000.0	169.0
15	0.0	10000.0	169.0
16	0.0	20000.0	169.0
17	0.0	30000.0	169.0
18	0.0	40000.0	169.0
19	0.0	50000.0	169.0
20	0.0	200.0	194.0
21	0.0	400.0	194.0
22	0.0	500.0	194.0
23	0.0	600.0	194.0
24	0.0	800.0	194.0

(continued)

Table 2.   (Continued)

Receptor Number	X Coordinate (m)	Y Coordinate (m)	Elevation (m)
25	0.0	1000.0	194.0
26	0.0	1250.0	194.0
27	0.0	1500.0	194.0
28	0.0	1750.0	194.0
29	0.0	2000.0	194.0
30	0.0	2500.0	194.0
31	0.0	3000.0	194.0
32	0.0	4000.0	194.0
33	0.0	5000.0	194.0
34	0.0	10000.0	194.0
35	0.0	20000.0	194.0
36	0.0	30000.0	194.0
37	0.0	40000.0	194.0
38	0.0	50000.0	194.0
39	0.0	200.0	220.0
40	0.0	400.0	220.0
41	0.0	500.0	220.0
42	0.0	600.0	220.0
43	0.0	800.0	220.0
44	0.0	1000.0	220.0
45	0.0	1250.0	220.0
46	0.0	1500.0	220.0
47	0.0	1750.0	220.0
48	0.0	2000.0	220.0
49	0.0	2500.0	220.0
50	0.0	3000.0	220.0
51	0.0	4000.0	220.0
52	0.0	5000.0	220.0
53	0.0	10000.0	220.0
54	0.0	20000.0	220.0
55	0.0	30000.0	220.0
56	0.0	40000.0	220.0
57	0.0	50000.0	220.0
58	0.0	200.0	270.0
59	0.0	400.0	270.0
60	0.0	500.0	270.0
61	0.0	600.0	270.0
62	0.0	800.0	270.0
63	0.0	1000.0	270.0
64	0.0	1250.0	270.0
65	0.0	1500.0	270.0
66	0.0	1750.0	270.0
67	0.0	2000.0	270.0
68	0.0	2500.0	270.0
69	0.0	3000.0	270.0
70	0.0	4000.0	270.0
71	0.0	5000.0	270.0
72	0.0	10000.0	270.0
73	0.0	20000.0	270.0
74	0.0	30000.0	270.0
75	0.0	40000.0	270.0
76	0.0	50000.0	270.0
77	0.0	200.0	470.0
78	0.0	400.0	470.0
79	0.0	500.0	470.0

Table 2.   (Continued)

Receptor Number	X Coordinate (m)	Y Coordinate (m)	Elevation (m)
80	0.0	600.0	470.0
81	0.0	800.0	470.0
82	0.0	1000.0	470.0
83	0.0	1250.0	470.0
84	0.0	1500.0	470.0
85	0.0	1750.0	470.0
86	0.0	2000.0	470.0
87	0.0	2500.0	470.0
88	0.0	3000.0	470.0
89	0.0	4000.0	470.0
90	0.0	5000.0	470.0
91	0.0	10000.0	470.0
92	0.0	20000.0	470.0
93	0.0	30000.0	470.0
94	0.0	40000.0	470.0
95	0.0	50000.0	470.0

Selected receptor locations analyzed by COMPDEP and
ISCST/LT were also analyzed by RTDMDEP. In order to
investigate the influence of terrain slope on model
predictions, three sets of RTDMDEP runs were made assuming
terrain slopes of 30.0°, 45.0°, and 60.0° respectively.

The coordinate systems used to express receptor locations
are defined as follows. Coordinate systems for both locations
are rectangular with the "+x" direction corresponding to east,
and the "+y" direction corresponding to north. Origins for
all systems correspond to the stack location. A complete
listing of receptor locations used by ISCST and COMPDEP is
given in Table 2.

Table 3.   Receptor Locations Used by RTDMDEP for
           Terrain Slopes of 30, 45, and 60°

Receptor Number	X Coordinate (m)	Y Coordinate (m)	Elevation (m)
1	0.0	500.0	194.0
2	0.0	500.0	220.0
3	0.0	500.0	270.0
4	0.0	500.0	470.0
5	0.0	1000.0	194.0
6	0.0	1000.0	220.0
7	0.0	1000.0	270.0
8	0.0	1000.0	470.0
9	0.0	5000.0	194.0
10	0.0	5000.0	220.0
11	0.0	5000.0	270.0
12	0.0	5000.0	470.0

Table 4a. COMPDEP Model Options

Building Wake Effects
Terrain Effects
Gradual Plume Rise
Buoyancy Induced Dispersion
Building Wake Effects
No Calm Processing
No Stack Downwash

Table 4b. RTDMDEP Model Options

Terrain Effects
Gradual Plume Rise
Buoyancy Induced Dispersion
No Stack Downwash
Pasquill-Gifford Dispersion Coefficients
Partial Plume Penetration (VPTG 0.0060 °K/m)
Unlimited Mixing for Stable Conditions
Plume Path Coefficient for Stability
    Classes 1-6: 0.50
Default Vertical Potential Temperature Gradients
    for Stability Classes 5 & 6
Stability Class-Dependent σ_y and σ_z
No Wind Direction Shear
Partial Plume Reflection
Off Centerline Horizontal Distribution Function for
    All Stabilities
Constant Emission Rate

Table 4c. ISCST and ISCLT Model Options

Terrain Effects
Rural Option
Gradual Plume Rise
Buoyancy Induced Dispersion
Building Wake Effects
No Stack Downwash
No Calm Processing
Default Vertical Potential Temperature Gradients

Receptor locations used by RTDMDEP for each of the three terrain slopes considered are found in Table 3.

Model Run Descriptions

   For each model, a concentration only run and a deposition run were made. The following input parameters were assumed by all model runs: exponents for power-law wind increase with height of 0.07, 0.07, 0.10, 0.15, 0.35, 0.55, anemometer height of 10.0 meters, surface roughness of 0.10 meters, no pollutant decay. Model options specific to each of the three

8

models used in the analysis are presented in Tables 4a - 4c.

Reflection Coefficients Used by ISCST

The algorithms used by COMPDEP and RTDMDEP to estimate pollutant deposition require an estimate of deposition velocity for each particle size analyzed. This is done automatically via the inclusion of the CARB (1986) algorithms. Unlike these models, ISCST requires that a reflection coefficient and gravitational settling velocity be input manually for each particle size. Values of these quantities used for all model runs were taken from U.S. EPA (1986b) and are listed in Table 5.

COMPARISON OF MODEL RESULTS

Calculated Air Concentrations

COMPDEP, ISCST and ISCLT Results: An examination of the results of Table 6. shows that for Receptors 1 - 19, which correspond to flat terrain, COMPDEP and ISCST predict essentially equivalent atmospheric pollutant concentrations. This agrees with anticipated results since the concentration algorithms used by both models are similar for conditions in which the effects of terrain and pollutant deposition are neglected. ISCLT and COMPDEP results differ, but only marginally. Excluding Receptor 1, for points within 5000.0 meters from the source, ISCLT results are about 1.1 times those of COMPDEP. At distances greater than and equal to 5000.0 meters, ISCLT concentrations are 0.9 times COMPDEP's.

Once terrain elevations are introduced, differences in model predictions increase. For receptors with elevations of half the stack height, Receptors 20 - 38, ISCST predictions exceed those of COMPDEP by a factor of 1.6 on an average basis and individually by as much as a factor of 5.2 (at Receptor 1) at distances less than or equal to 10 km from the source. At distances greater than 10 km, both models again predict nearly identical concentrations. As with Receptors 1 - 19, ISCLT concentrations are greater on average than COMDEP's (by a factor of 1.7) for points near the source (less than 10000.0 meters), and less than COMPDEP's (by a factor of 0.9) at larger downwind distances. As with ISCST, the maximum deviation between ISCLT results and those of COMPDEP occurs 200 meters from the source (Receptor 1) where ISCLT overpredicts COMPDEP by a factor of 5.0.

Table 5. Reflection Coefficients and Gravitational
          Settling Velocities

Particle Diameter (μm)	Reflection Coefficient	Settling Velocity (cm/s)
$\leq$ 2.0	0.961	5.55E-03
2.0 - 10.0	0.861	2.55E-01
$\geq$ 10.0	0.714	2.22E+00

TABLE 6.   AVERAGE ATMOSPHERIC CONCENTRATIONS,
           NEGLECTING DEPOSITION EFFECTS, PREDICTED
           BY COMPDEP, ISCST and ISCLT

Receptor Number	Concentrations ($\mu g/m^3$) COMPDEP	ISCST	ISCLT
1	0.848E-02	0.849E-02	0.756E-02
2	0.112E+00	0.112E+00	0.137E+00
3	0.165E+00	0.165E+00	0.208E+00
4	0.212E+00	0.212E+00	0.261E+00
5	0.295E+00	0.296E+00	0.336E+00
6	0.350E+00	0.350E+00	0.379E+00
7	0.355E+00	0.355E+00	0.375E+00
8	0.342E+00	0.342E+00	0.357E+00
9	0.322E+00	0.322E+00	0.333E+00
10	0.300E+00	0.300E+00	0.309E+00
11	0.258E+00	0.258E+00	0.263E+00
12	0.223E+00	0.223E+00	0.226E+00
13	0.171E+00	0.171E+00	0.171E+00
14	0.137E+00	0.137E+00	0.135E+00
15	0.640E-01	0.640E-01	0.605E-01
16	0.284E-01	0.284E-01	0.256E-01
17	0.174E-01	0.174E-01	0.153E-01
18	0.124E-01	0.124E-01	0.106E-01
19	0.949E-02	0.949E-02	0.804E-02
20	0.404E-01	0.212E+00	0.203E+00
21	0.271E+00	0.829E+00	0.856E+00
22	0.381E+00	0.101E+01	0.103E+01
23	0.474E+00	0.109E+01	0.111E+01
24	0.589E+00	0.107E+01	0.110E+01
25	0.622E+00	0.965E+00	0.993E+00
26	0.581E+00	0.814E+00	0.837E+00
27	0.532E+00	0.694E+00	0.713E+00
28	0.484E+00	0.602E+00	0.617E+00
29	0.441E+00	0.528E+00	0.541E+00
30	0.367E+00	0.419E+00	0.426E+00
31	0.310E+00	0.343E+00	0.348E+00
32	0.231E+00	0.248E+00	0.248E+00
33	0.181E+00	0.191E+00	0.189E+00
34	0.795E-01	0.809E-01	0.769E-01
35	0.332E-01	0.334E-01	0.304E-01
36	0.199E-01	0.199E-01	0.176E-01
37	0.139E-01	0.139E-01	0.121E-01
38	0.106E-01	0.106E-01	0.902E-02
39	0.117E+01	0.128E+02	0.110E+02
40	0.132E+01	0.650E+01	0.640E+01
41	0.148E+01	0.500E+01	0.505E+01
42	0.156E+01	0.401E+01	0.411E+01
43	0.153E+01	0.282E+01	0.293E+01
44	0.139E+01	0.212E+01	0.222E+01
45	0.118E+01	0.161E+01	0.169E+01
46	0.101E+01	0.129E+01	0.134E+01
47	0.870E+00	0.106E+01	0.110E+01
48	0.757E+00	0.890E+00	0.917E+00
49	0.589E+00	0.663E+00	0.676E+00
50	0.473E+00	0.518E+00	0.524E+00
51	0.330E+00	0.350E+00	0.349E+00
52	0.247E+00	0.259E+00	0.254E+00
53	0.966E-01	0.981E-01	0.928E-01

Table 6.   (Continued)

Receptor	Concentrations (μg/m^3)		
Number	COMPDEP	ISCST	ISCLT
54	0.377E-01	0.379E-01	0.344E-01
55	0.219E-01	0.220E-01	0.194E-01
56	0.152E-01	0.152E-01	0.132E-01
57	0.114E-01	0.114E-01	0.975E-02
58	0.851E+01	0.128E+02	0.110E+02
59	0.569E+01	0.650E+01	0.640E+01
60	0.489E+01	0.500E+01	0.505E+01
61	0.422E+01	0.401E+01	0.411E+01
62	0.320E+01	0.282E+01	0.293E+01
63	0.248E+01	0.212E+01	0.222E+01
64	0.189E+01	0.161E+01	0.169E+01
65	0.149E+01	0.129E+01	0.134E+01
66	0.121E+01	0.106E+01	0.110E+01
67	0.100E+01	0.890E+00	0.917E+00
68	0.733E+00	0.663E+00	0.676E+00
69	0.564E+00	0.518E+00	0.524E+00
70	0.374E+00	0.350E+00	0.349E+00
71	0.271E+00	0.258E+00	0.254E+00
72	0.992E-01	0.981E-01	0.928E-01
73	0.374E-01	0.379E-01	0.344E-01
74	0.216E-01	0.220E-01	0.194E-01
75	0.148E-01	0.152E-01	0.132E-01
76	0.111E-01	0.114E-01	0.975E-02
77	0.403E+01	0.128E+02	0.110E+02
78	0.270E+01	0.650E+01	0.640E+01
79	0.231E+01	0.500E+01	0.505E+01
80	0.199E+01	0.401E+01	0.411E+01
81	0.151E+01	0.282E+01	0.293E+01
82	0.117E+01	0.212E+01	0.222E+01
83	0.896E+00	0.161E+01	0.169E+01
84	0.702E+00	0.129E+01	0.134E+01
85	0.570E+00	0.106E+01	0.110E+01
86	0.473E+00	0.890E+00	0.917E+00
87	0.346E+00	0.663E+00	0.676E+00
88	0.266E+00	0.518E+00	0.524E+00
89	0.176E+00	0.350E+00	0.349E+00
90	0.128E+00	0.258E+00	0.254E+00
91	0.468E-01	0.981E-01	0.925E-01
92	0.177E-01	0.379E-01	0.344E-01
93	0.102E-01	0.220E-01	0.194E-01
94	0.702E-02	0.152E-01	0.132E-01
95	0.527E-02	0.114E-01	0.975E-02

A similar pattern is seen for receptors which are equal to the stack height in elevation (Receptors 39 - 57).   ISCST predictions for receptors within 10 km of the stack exceed those of COMPDEP by an average factor of 1.7.   The maximum deviation between the two models' predictions occurs 200 meters from the source where the ISCST concentration is a factor of 10.9 times that of COMPDEP.   As with the previous two groups of receptors, concentration predictions are essentially identical for distant points (those in excess of 10 km from the source).   ISCLT results are a maximum of 9.4

times those of COMPDEP and average about 1.8 times higher
within 10 km of the source. At distances greater than 10 km,
ISCLT concentrations again average 0.9 times those of COMPDEP
and are a maximum of 1.2 times greater at Receptor 57. In
summary, for the elevations considered so far, it is apparent
that while average concentrations predicted by ISCST, ISCLT,
and COMPDEP agree rather well even in complex terrain,
individual differences between ISCST/LT and COMPDEP differ by
as much as a factor of ten.

An examination of the remaining two receptor groups, those
which differ from those for the first three groups. While
concentrations predicted for receptors 100 meters above stack
height and less than 600 meters from the source are 1.0 to 1.5
times higher than COMPDEP's, ISCST predictions are less than
COMPDEP's between 600 and 20000 meters from the stack by an
average factor of 0.9. This discrepancy may be explained by
the fact that for terrain above the stack, ISCST substitutes
the stack height for the actual terrain elevation in
concentration calculations. Consequently, this approach
yields almost random values which should be considered highly
uncertain. Concentrations predicted by COMPDEP are greater at
this elevation than at stack height. This indicates that
COMPDEP's terrain correction factor has little effect at this
elevation as the plume height (the sum of stack height and
plume rise) is still above terrain elevation. ISCST
concentrations beyond 20 km are again equal to those of
COMPDEP. ISCLT concentrations are also about 1.2 times those
of COMPDEP at distances less than 600 meters and 0.9 times
less than COMPDEP's at greater downwind distances.

In contrast to results for receptors 100 meters above stack
elevation, concentrations predicted by both ISCST and ISCLT at
elevations 300 meters above stack top are greater on average
than those of COMPDEP by factors of 2.1 and 2.0, respectively,
for all locations. The maximum deviation occurs at Receptor
77 where ISCST and ISCLT overpredict COMPDEP by factors of 3.2
and 2.7. While concentrations predicted by ISCST and ISCLT
are identical to those for the previous two groups, those
calculated by COMPDEP are significantly less than at the
previous elevation. Thus, for this higher elevation,
COMPDEP's terrain correction factor has a considerable
influence on model predictions.

RTDMDEP Results: An examination of concentrations
predicted by the third model used in the analysis, RTDMDEP,
provides additional interesting results (See Table 7.). As
previously stated, RTDMDEP employs a partial plume penetration
algorithm which, for elevated terrain, tends to prevent
erroneous magnification of pollutant concentrations. It is
thus expected that concentrations predicted by RTDMDEP should
be less than those of COMPDEP for elevated terrain. For the
most part this expectation is met. At Receptor 2, processes
that occur close to the source such as building wake effects,
tend to dominate pollutant concentrations. Excluding this
point, values predicted by COMPDEP for receptors equal to and
above stack height exceed those of RTDMDEP for all terrain
slope values. For receptors at or above stack height COMPDEP
predictions exceed those of RTDMDEP by as little as a factor
of 1.1 at Receptor 6 and by as much as 2.0 at Receptor 3.
This comparison is consistent with previous evaluations of
COMPLEX I and RTDM (Paine and Egan, 1986) which shows that

12

Table 7.   AVERAGE ATMOSPHERIC CONCENTRATIONS, NEGLECTING
           DEPOSITION EFFECTS, PREDICTED BY RTDMDEP, COMPDEP,
           ISCST and ISCLT

		Concentrations ($\mu g/m^3$)			
	RTDMDEP				
Rec. No.	30° Slope	45° Slope	60° Slope	COMPDEP	ISCST	ISCLT
1	0.40E+00	0.40E+00	0.40E+00	0.38E+00	0.10E+01	0.10E+01
2	0.20E+01	0.20E+01	0.22E+01	0.15E+01	0.50E+01	0.50E+01
3	0.24E+01	0.24E+01	0.24E+01	0.49E+01	0.50E+01	0.50E+01
4	0.18E+01	0.18E+01	0.17E+01	0.23E+01	0.50E+01	0.50E+01
5	0.66E+00	0.66E+00	0.66E+00	0.62E+00	0.96E+00	0.99E+01
6	0.13E+01	0.13E+01	0.13E+01	0.14E+01	0.21E+01	0.22E+01
7	0.14E+01	0.14E+01	0.14E+01	0.25E+01	0.21E+01	0.22E+01
8	0.11E+01	0.10E+01	0.99E+00	0.12E+01	0.21E+01	0.22E+01
9	0.18E+00	0.18E+00	0.18E+00	0.18E+00	0.19E+00	0.19E+00
10	0.17E+00	0.18E+00	0.18E+00	0.25E+00	0.26E+00	0.25E+00
11	0.17E+00	0.17E+00	0.17E+00	0.27E+00	0.26E+00	0.25E+00
12	0.11E+00	0.11E+00	0.11E+00	0.13E+00	0.26E+00	0.25E+00

both models have a tendency to overestimate pollutant
concentrations in complex terrain and the atmospheric
concentrations predicted by RTDM are less than those
determined by COMPLEX I.

For Receptors 1 and 5, which are lower than stack height
and near the source, the COMPDEP concentrations are less than
those from RTDMDEP. This variation is most likely due to the
fact that the partial plume reflection algorithm of RTDMDEP
does not affect terrain below stack elevation. Of further
interest, is the fact that for all receptors above stack
height, differences in values predicted by the two models
decrease with increasing receptor elevation. This, and the
fact that the algorithms RTDMDEP uses to compute terrain
effects are much more sophisticated than those in COMPDEP,
suggests that the simple terrain correction factor used by
COMPDEP works reasonably well for the range of elevations
considered.

Variation of terrain slope seems to have a minimal effect
on concentrations predicted by RTDMDEP for the conditions
modeled. The greatest differences are seen at locations of
large downwind distances and high elevations (Receptors 8, 10,
11, and 12,). The highest change occurs at Receptor 8 where a
13.9 percent decrease in concentration occurs as terrain slope
varies from 30° to 60°. It is possible that additional
variation in hill height used in model calculations could
magnify the differences seen.

For all locations, concentrations predicted by ISCST and
ISCLT are greater than those of RTDMDEP. Both ISCST and ISCLT
predict concentrations which range from 1.1 to 2.8 times those
of RTDMDEP.

Calculated Values of Pollutant Deposition

COMPDEP, ISCST and ISCLT Results: Unlike atmospheric
concentration calculations which were essentially identical,
predicted values for dry deposition made by COMPDEP (Table 8.)
are different than those for ISCST and ISCLT in flat terrain
(Receptors 1 - 19). Deposition rates for receptors within
3000 meters of the source predicted by ISCST and ISCLT are on
average, 2.3 and 2.9 times higher than those of COMPDEP.
Maximum deposition rates predicted by the two models are,
respectively, as much as 10.0 and 9.5 times those of COMPDEP.
Beyond this distance, however, the opposite is true as
deposition rates predicted by COMPDEP are 1.1 to 4.2 times
those of ISCST and 1.1 to 4.7 times those estimated by ISCLT.
Examination of the remaining receptors yields similar results.
For points near the source, ISCST deposition rates vary from
1.0 to 37.6 and average 3.0 times those of COMPDEP. ISCLT
deposition rates average 3.5 and range from 1.1 to 39.3 times
COMPDEP's. As downwind distance increases, values predicted
by COMPDEP average 2.5 times higher than those of ISCST and
2.4 times greater than ISCLT's. Furthermore, the point at
which COMPDEP predictions exceed those of ISCST/LT shifts
toward the stack as terrain height increases for Receptors 25,
50, and 100 meters above stack height. The fact that this
trend is not seen for the receptors 300 meters above the stack
is probably due to the increasing influence of the terrain
correction factor used by COMPDEP.

Table 8.   AVERAGE ANNUAL DRY DEPOSITION PREDICTED
           BY COMPDEP AND ISCST

Receptor Number	Deposition ($g/m^2/y$)		
	COMPDEP	ISCST	ISCLT
1	0.816E-03	0.814E-02	0.778E-02
2	0.108E-01	0.626E-01	0.743E-01
3	0.158E-01	0.782E-01	0.936E-01
4	0.203E-01	0.872E-01	0.101E+00
5	0.279E-01	0.956E-01	0.105E+00
6	0.322E-01	0.917E-01	0.986E-01
7	0.318E-01	0.703E-01	0.752E-01
8	0.300E-01	0.562E-01	0.598E-01
9	0.277E-01	0.454E-01	0.480E-01
10	0.254E-01	0.371E-01	0.390E-01
11	0.212E-01	0.256E-01	0.268E-01
12	0.180E-01	0.185E-01	0.185E-01
13	0.134E-01	0.105E-01	0.107E-01
14	0.104E-01	0.693E-02	0.691E-02
15	0.420E-02	0.188E-02	0.178E-02
16	0.145E-02	0.475E-03	0.438E-03
17	0.770E-03	0.225E-03	0.200E-03
18	0.500E-03	0.139E-03	0.124E-03
19	0.359E-03	0.983E-04	0.870E-04
20	0.404E-02	0.150E+00	0.150E+00
21	0.267E-01	0.335E+00	0.349E+00
22	0.372E-01	0.331E+00	0.347E+00
23	0.455E-01	0.297E+00	0.314E+00

Table 8.   (Continued)

Receptor Number	Deposition (g/m^2/y)		
	COMPDEP	ISCST	ISCLT
24	0.545E-01	0.217E+00	0.233E+00
25	0.555E-01	0.156E+00	0.168E+00
26	0.505E-01	0.981E-01	0.106E+00
27	0.451E-01	0.699E-01	0.756E-01
28	0.402E-01	0.522E-01	0.562E-01
29	0.360E-01	0.404E-01	0.433E-01
30	0.291E-01	0.257E-01	0.275E-01
31	0.239E-01	0.177E-01	0.181E-01
32	0.170E-01	0.962E-02	0.996E-02
33	0.127E-01	0.611E-02	0.620E-02
34	0.460E-02	0.156E-02	0.148E-02
35	0.153E-02	0.402E-03	0.361E-03
36	0.806E-03	0.197E-03	0.170E-03
37	0.520E-03	0.125E-03	0.110E-03
38	0.369E-03	0.899E-04	0.790E-04
39	0.916E-01	0.344E+01	0.360E+01
40	0.119E+00	0.908E+00	0.106E+01
41	0.132E+00	0.574E+00	0.679E+00
42	0.137E+00	0.394E+00	0.468E+00
43	0.129E+00	0.215E+00	0.256E+00
44	0.113E+00	0.134E+00	0.158E+00
45	0.926E-01	0.787E-01	0.923E-01
46	0.769E-01	0.533E-01	0.619E-01
47	0.646E-01	0.383E-01	0.439E-01
48	0.549E-01	0.287E-01	0.326E-01
49	0.411E-01	0.173E-01	0.194E-01
52	0.318E-01	0.115E-01	0.134E-01
51	0.209E-01	0.613E-02	0.648E-02
52	0.148E-01	0.386E-02	0.394E-02
53	0.485E-02	0.105E-02	0.956E-03
54	0.157E-02	0.307E-03	0.259E-03
55	0.816E-03	0.161E-03	0.134E-03
56	0.520E-03	0.108E-03	0.920E-04
57	0.367E-03	0.798E-04	0.680E-04
58	0.535E+00	0.344E+01	0.360E+01
59	0.384E+00	0.908E+00	0.106E+01
60	0.334E+00	0.574E+00	0.679E+00
61	0.288E+00	0.394E+00	0.468E+00
62	0.216E+00	0.215E+00	0.256E+00
63	0.164E+00	0.134E+00	0.158E+00
64	0.122E+00	0.787E-01	0.923E-01
65	0.944E-01	0.533E-01	0.619E-01
66	0.752E-01	0.383E-01	0.439E-01
67	0.613E-01	0.287E-01	0.326E-01
68	0.434E-01	0.173E-01	0.194E-01
69	0.325E-01	0.115E-01	0.134E-01
70	0.204E-01	0.613E-02	0.648E-02
71	0.142E-01	0.386E-02	0.394E-02
72	0.449E-02	0.105E-02	0.956E-03
73	0.144E-02	0.307E-03	0.259E-03
74	0.744E-03	0.161E-03	0.134E-03
75	0.472E-03	0.108E-03	0.920E-04
76	0.333E-03	0.798E-04	0.680E-04

(continued)

Table 8. (Continued)

Receptor Number	Deposition ($g/m^2/y$) COMPDEP	ISCST	ISCLT
77	0.252E+00	0.344E+01	0.360E+01
78	0.181E+00	0.908E+00	0.106E+01
79	0.156E+00	0.574E+00	0.679E+00
80	0.134E+00	0.394E+00	0.468E+00
81	0.101E+00	0.215E+00	0.256E+00
82	0.768E-01	0.134E+00	0.158E+00
83	0.571E-01	0.787E-01	0.923E-01
84	0.441E-01	0.533E-01	0.619E-01
85	0.351E-01	0.383E-01	0.439E-01
86	0.286E-01	0.287E-01	0.326E-01
87	0.203E-01	0.173E-01	0.194E-01
88	0.151E-01	0.115E-01	0.134E-01
89	0.953E-02	0.613E-02	0.648E-02
90	0.661E-02	0.386E-02	0.394E-02
91	0.209E-02	0.105E-02	0.956E-03
92	0.671E-03	0.307E-03	0.259E-03
93	0.348E-03	0.161E-03	0.134E-03
94	0.221E-03	0.108E-03	0.920E-04
95	0.156E-03	0.798E-04	0.680E-04

RTDMDEP Results: Deposition rates were also calculated by RTDMDEP (Table 9.) for terrain 25, 50, 100, and 300 meters above stack height. Differences in deposition rates predicted by the two models are less than for concentrations. COMPDEP predictions range from 0.8 to 1.9 times those of the RTDMDEP. Like COMPDEP, RTDMDEP predicted deposition rates 2.7 times lower on average and as much as 8.7 times lower than those of ISCST near the source. For these same points, ISCLT results were 3.1 times higher, on average, and as much as 9.1 times RTDMDEP's. At larger downwind distances, RTDMDEP values were 1.9 to 3.1 times higher than those of ISCST and ISCLT.

CONCLUSIONS

One of the goals of this analysis was to determine whether it is appropriate to use flat terrain models such as ISCST and ISCLT when modeling pollutant transport in complex terrain. Comparison of the flat terrain models, ISCST and ISCLT, with the modified complex terrain models, COMPDEP and RTDMDEP, showed, on average, that ISCST and ISCLT generally predicted marginally higher annual average atmospheric pollutant concentrations which were roughly twice those of COMPDEP and three times those of RTDMDEP. A few isolated cases existed in which ISCST and ISCLT actually gave slightly unconservative predictions with respect to those made by COMPDEP and RTDMDEP. These occurred primarily at large downwind distances and at receptor elevations of 100.0 meters above stack elevation. Many risk assessments, such as those weighted according to population density, are more a function of overall average pollutant concentration near the source than a maximum concentration at a single point. Based on the results of this

Table 9.   AVERAGE ANNUAL DRY DEPOSITION PREDICTED BY RTDMDEP,
           COMPDEP, ISCST, AND ISCLT

Deposition (g/m^2/y)

Rec. No.	RTDMDEP 30° Slope	RTDMDEP 45° Slope	RTDMDEP 60° Slope	COMPDEP	ISCST	ISCLT
1	0.38E-01	0.38E-01	0.38E-01	0.37E-01	0.33E+00	0.35E+00
2	0.16E+00	0.16E+00	0.18E+00	0.13E+00	0.57E+00	0.68E+00
3	0.18E+00	0.18E+00	0.18E+00	0.33E+00	0.57E+00	0.68E+00
4	0.16E+00	0.16E+00	0.15E+00	0.16E+00	0.57E+00	0.68E+00
5	0.58E-01	0.58E-01	0.58E-01	0.55E-01	0.16E+00	0.17E+00
6	0.10E+00	0.10E+00	0.10E+00	0.11E+00	0.13E+00	0.16E+00
7	0.97E-01	0.97E-01	0.97E-01	0.16E+00	0.13E+00	0.16E+00
8	0.95E-01	0.84E-01	0.79E-01	0.77E-01	0.13E+00	0.16E+00
9	0.13E-01	0.13E-01	0.13E-01	0.13E-01	0.61E-02	0.62E-02
10	0.12E-01	0.12E-01	0.12E-01	0.15E-01	0.39E-02	0.39E-02
11	0.10E-01	0.10E-01	0.10E-01	0.14E-01	0.39E-02	0.39E-02
12	0.73E-02	0.74E-02	0.75E-02	0.66E-02	0.39E-02	0.39E-02

analysis, use of ISCST or ISCLT in these risk assessments
would be expected to give results that would range from about
0.9 to 3.0 times those obtained if either COMPDEP or RTDMDEP
were used. Hence, for such assessments, use of ISCST or ISCLT
in complex terrain appears to be valid.

Somewhat different conclusions can be drawn when
considering maximum point concentrations. Whereas differences
in average concentrations predicted by flat and complex
terrain models were small, differences in maximum point
concentrations were much more significant, differing in some
instances by an order of magnitude. In all cases, the maximum
point concentrations were predicted by ISCST and ISCLT. Thus
in some cases, such as for risk assessments which base their
exposures on maximum point concentrations of pollutant, use of
ISCST and ISCLT in complex terrain must be questioned. If the
user can tolerate a factor of ten conservatism, these models
can be applied. However, if this is not the case,
consideration should be given to either COMPDEP or RTDMDEP.
Based on the experience gained during this study, COMPDEP is
recommended over RTDMDEP for use in these situations. COMPDEP
is easier to use, faster, and can be applied to more physical
scenarios than RTDMDEP. Only for cases in which
concentrations estimated by COMPDEP appear to be too
conservative is RTDMEP recommended for use.

It should be noted that in spite of their technical
improvements, currently recommended complex terrain screening
models (U.S. EPA, 1986a) are still rather limited in their
ability to model atmospheric transport in complex terrain.
Because of their Gaussian plume formulation, their treatment
of terrain is limited to a few algorithms which modify plume
height. In the future it is possible that a refined complex
terrain model, capable of realistically simulating flow in
complex terrain, will be developed which can be used to
generate estimates of annual average atmospheric pollutant
concentrations.

The second main objective of this study was to determine whether the use of improved deposition algorithms would result in better estimates of pollutant deposition rate. The deposition algorithms incorporated into COMPDEP and RTDMDEP are, in theory, more advanced than those contained in ISCST/LT and should provide more realistic estimates of pollutant deposition rate. In addition, unlike ISCST/LT which require the user input highly subjective values for reflection coefficient, these models offer a systematic and automated means by which deposition velocities can be estimated. However, accurate conclusions regarding which provide the best estimates of dry deposition velocity cannot be obtained without reference to quantitative evaluations between predicted and measured values. Unfortunately such evaluations are rare. Bowers et al. (1981) found that ISCST/LT produced results within a factor of two of measured values. In some instances calculated values were lower than measured values. Based on these results, deposition rates predicted COMPDEP and RTDMDEP which are as much as 39 times lower than those of ISCST/LT appear to grossly underestimate pollutant deposition rates. However, the deposition velocities from Bowers et al. (1981) were calculated under a limited set of atmospheric conditions and downwind distances and for particle sizes which ranged from 50 to 200 microns, much larger than those addressed in this analysis. In contrast to these results Tesche et al. (1987) indicated that under certain conditions ISCST can significantly overestimate dry deposition. Although no quantitative estimates of the degree of overprediction by ISCST were reported, the Tesche et al. (1987) results indicate that COMPDEP and RTDMDEP may provide better estimates of deposition rate under certain conditions. Thus, current evaluations of ISCST/LT performance in estimating deposition rate appear contradictory. Until more detailed evaluations of the performance of models such as ISCST/LT in estimating deposition rates are available, meaningful recommendations regarding whether the deposition algorithms developed for this study are indeed superior to those found in ISCST/LT cannot be made.

REFERENCES

Bowers, J. F., Anderson, A. J., 1981, An Evaluation Study for the Industrial Source Complex (ISC) Dispersion Model, EPA-450/4-81-002.

California Air Resources Board (CARB), 1986, Subroutines for calculating dry deposition velocities using Sehmel's curves.

Environmental Research and Technology (ERT), 1987, User's Guide to the Rough Terrain Diffusion Model (RTDM), Rev 3.20, ERT Document No. P-D535-585, Concord, Ma.

Gale Research Co, 1983, "Climate Normals for the U.S. (Base: 1951-80)," Book Tower, Detroit.

Holzworth, G. C., 1972, Mixing Heights, Wind Speeds, and Potential for Urban Air Pollution Throughout the Contiguous United States, RTP.

Paine, R. J., Egan, B. A., 1986, Results of Additional
     Evaluation of the Rough Terrain Diffusion Model (RTDM),
     for presentation at the 79th Annual Meeting of the Air
     Pollution Control Association (APCA), Minneapolis.

Sehmel, G. A., 1980, Particle and gas dry deposition: A
     review, Atmos. Env. 14: 983-1011.

Tesche, T. W., Kapahi, R., Honrath, R., Dietrich, W., 1987,
     Improved Dry Deposition Estimates for Air Toxics Risk
     Assessment, for presentation at the 80th annual meeting of
     the Air Pollution Control Association (APCA), New York.

Travis, C. C., Yambert, M. W., 1989, An Evaluation of
     Models Used to Compute Particulate Dry Deposition Velocity
     to Pasture, Agricultural Land, and Similar Surfaces, in:
     "Municipal Waste Incineration Risk Assessment," Plenum,
     New York.

Turner, D. B., 1986, Fortran computer code/user's guide for
     COMPLEX I Version 86064: An Air Quality Dispersion Model,
     in Section 4. Additional Models for Regulatory Use. Source
     File 31 Contained in UNAMAP (VERSION 6), NTIS PB 96-222
     361/AS.

U.S. Environmental Protection Agency, 1979, Industrial
     Source Complex (ISC) Dispersion Model User's Guide, Vol. I,
     EPA-450/4-79-030.

Rao, K. Shankar, Satterfield, Lynne, 1982, MPTER-DS: The
     MPTER Model Including Deposition and Sedimentation,
     EPA-600/8-82-024.

U.S. Environmental Protection Agency, 1986a, Guideline on
     Air Quality Models (Revised), EPA-450/2-78-072R.

U.S. Environmental Protection Agency, 1986b, Air Quality
     Modeling Analysis of Municipal Waste Combustors. Report
     prepared by PEI Associates, Cincinnati, Ohio and H. E.
     Cramer Company, Salt Lake City, Utah.

U.S. Environmental Protection Agency, 1987, Air Dispersion
     Modeling of a Municipal Waste Combustor in Rutland,
     Vermont. Report prepared by PEI Associates, Cincinnati,
     Ohio.

SMALL PARTICLE DEPOSITION

IN AIR QUALITY MODELING

Ray Kapahi

EMCON Associates
1433 North Market Boulevard
Sacramento, California 95864

INTRODUCTION

The combustion of municipal waste can potentially result in the release of toxic air contaminants into the atmosphere. Even though stringent air pollution controls are used in waste-to-energy facilities, some toxic materials can still be released into the atmosphere in the form of gases and fine particulates. This paper focuses on modeling the atmospheric dispersion and dry deposition of fine particulates.

For present discussion, "fine particulates" are defined as solid particles that have an aerodynamic diameter of roughly 10 to 20 micrometers, or smaller. This size range is particularly relevant to municipal WTE facilities since best available control technology, such as fabric filters, effectively remove any particulates above this size range. These particulates are characterized by having relatively low gravitational settling velocities as compared to the horizontal ambient wind speed. In some ways fine particulates behave similar to gaseous pollutants. As such, they can enter the human body via inhalation and/or the terrestrial pathways.

Figure 1 illustrates the linkage between the emissions source, the dispersion and dry deposition models, and the food chain. Since risk to human health is related to the ambient concentration and the dry deposition rates of particulates, any over- or underestimates of these quantities will result in corresponding under- or overestimates of potential risks to human health.

Currently, there are few models available to the modeling community that can realistically estimate the ambient concentrations of fine particulates and are acceptable to the regulatory agencies. This is partly because most air quality standards were based on ambient concentration rather than on dry deposition rates. As a result, many air quality models have been developed over the last three decades. In addition, there have been model evaluation performance of air quality models. In comparison, there has been a relatively low level of activity in developing deposition models or field studies in establishing their performance.

This paper describes the development of two models for improving estimates of fine particulate dispersion and dry deposition. The first is based on adapting an existing Gaussian plume model so that key dispersion/dry deposition processes can be treated in a physically realistic manner. The second model is similar to the first, except it can be used in complex terrain.

We begin by briefly reviewing several existing deposition models and their applications to health risk assessments.

Municipal Waste Incineration Risk Assessment
Edited by C.C. Travis, Plenum Press, New York, 1991

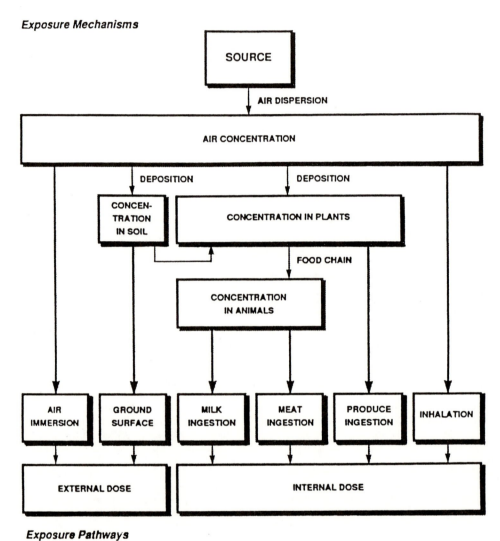

Figure 1. Pathways of airborne contaminants into the environment.

We begin by briefly reviewing several existing deposition models and their applications to health risk assessments.

<u>A Brief Review of Existing Models</u>

Two key requirements to be met by air quality models suitable for use in estimating ambient concentrations and dry deposition rates are:

1) Conserve Mass of Particulates; and

2) Simulate the Turbulent, Gravitational and Brownian Transfer of Particulate Mass from Plume to Ground.

A review of the literature (Hanna et al 1982, Horst 1979, and others) indicates three basic approaches in modeling particulates:

1) Source Depletion Models;

2) Surface Depletion Models; and

3) Gradient Transfer Models.

Source depletion models represent the simplest approach for modeling particulates. This approach adjusts the source strength to account for deposition of particulates. The problem with this approach is that the depleted mass of particulates is distributed along the entire vertical extent of the plume, not just the portion of the plume adjacent to the surface. The Industrial Source Complex Model (ISC) recommended by the Environmental Protection Agency (EPA 1984) is an example of a source depletion model. This model is designed for large particles (50 micrometers or greater) that have appreciable settling velocities.

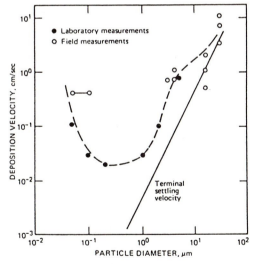

Figure 2. Laboratory and field measurements of deposition speeds of particulates and gases (after McMahon and Denison, 1979).

Surface depletion model (Horst 1974) corrects this problem by the use of "negative plumes" that simulate plume sedimentation and depletion. This approach is physically more realistic than the source depletion models. This, however, is at the expense of substantially increased computational requirements. Horst (1974), Prahm and Berkowicz (1978), and Yamartino (1981) have used this approach, but for the most part, surface depletion models have yet to find widespread usage.

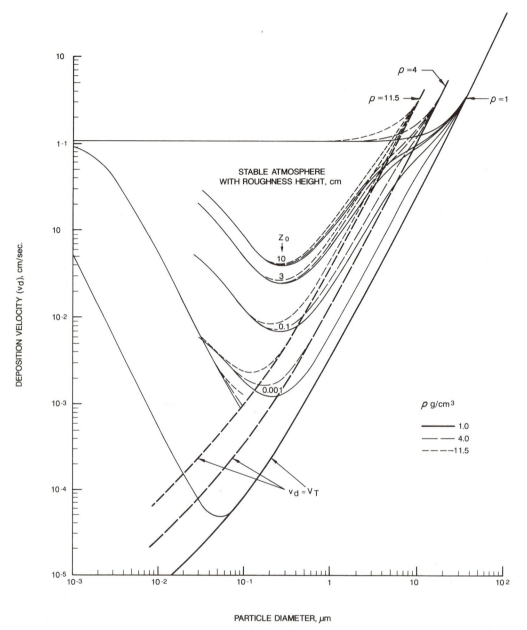

Figure 3. Predicted deposition velocities for particle densities of 1, 4, and 11.5 gm/cc (after Sehmel, 1980).

Rao (1986) has used the gradient transfer theory in simulating particulate dispersion and deposition. He has adapted several existing Gaussian based models in the UNAMAP suite of codes for use in dispersion/deposition calculations. Specifically, he modified the Texas Episodic Model (TACB 1979) and the MPTER model for use in estimating short and long-term estimates respectively. These models treat both dry deposition, gravitational settling and first order chemical transformation. They provide simultaneous estimates of ambient concentrations and dry deposition rates at each receptor.

The difficulty with all of the above models is that the deposition velocities must be specified by the user, yet little guidance is provided on how they are to be estimated. For fine particulates, the deposition velocity is related not just to particulate size and density, but also to atmospheric turbulence. Since the latter changes with time, it would be desirable to include a model that simulates the variation of deposition velocity with particle size, density and atmospheric turbulence. In the next section, we describe how several existing models can be improved to include this variation of deposition velocity.

Development of Improved Gaussian Based Deposition Models

As noted above, one immediate improvement to existing models is to incorporate algorithms that estimate deposition velocity as a function of particle size, particle density, and atmospheric turbulence.

Measurements by Sehmel (1980) and McMahon and Denison (1979) indicate that the overall deposition velocity is greater than the gravitational settling velocity (see Figure 2). For particles in the 0.1 to 1 micrometer range, the deposition velocity can be over an order of magnitude greater the settling velocity. Clearly, if one were to assume deposition velocity is solely as a result of gravity, then one would substantially underestimate the actual deposition rates and overestimate the ambient concentration.

Sehmel and Hodgson (1974) have developed a model for estimating dry deposition velocities as a function of particle size, particle density, underlying surface and friction velocity (a measure of atmospheric turbulence or stability). Deposition velocities predicted by this model are shown in Figure 3.

The first model, the Radian Deposition Model (RDM), (Kapahi 1988a) is based on the PEM-4 model developed by Rao (1984). PEM-4 is an urban model for use with multiple point and area sources located in flat terrain. It is, however, an episodic (short-term) model and cannot be used to estimate annual ground level concentrations and deposition rates. These are often needed for estimating long-term or lifetime health risks. In addition, PEM-4 requires the user to specify deposition and settling velocities.

RDM overcomes these limitations by incorporating the deposition velocity model of Sehmel and Hodgson (1974) and extends the averaging time from 24 hours to 1 year.

A second Gaussian based deposition model, MPTER-DS, was also modified to incorporate the Sehmel and Hodgson (1977) deposition velocity model. This model, called MPT (Kapahi 1986) is similar to the RDM model, except it can be used in complex terrain in both urban and rural areas.

Both models accept standard National Climatic Data Center (NCDC) hourly meteorological data and can be readily used to estimate concentrations and dry deposition rates for averaging periods form 1 hour to 1 year. A comparison of these models results with field measurements are currently underway and will be presented in a subsequent study (Kapahi 1989b).

References

EPA, 1984. Guidelines on Air Quality Models (Revised), USEPA, Research Triangle Park, N.C.

Hanna, S.R., et al, 1982. Handbook of Atmospheric Diffusion, ATDL, National Oceanic and Atmospheric Administration. Published by Technical Information Center, U. S. Department of Energy.

Horst, T.W., 1977. "A Surface Depletion Model for Deposition from a Gaussian Plume," Proc. of Atmosphere-Surface Exchange of Particulate and Gaseous Pollutants (1974), September 4-6, 1974.

Horst, T.W., 1979. "A Review of Gaussian Diffusion-Deposition Models," in D.S. Shriner, et al, Atmospheric Sulfur Deposition: Environmental Impact and Health Effects, Ann Arbor Science Publishers, Ann Arbor, MI.

Kapahi, R., 1988a. Radian Deposition Model (RDM), Radian Corporation, Sacramento, CA.

Kapahi, R., 1989b. "MPT: A Model for Estimating Dispersion and Deposition in Complex Terrain," Ecoserve Environmental Services, Pittsburg, CA.

McMahon, T.A., and P.J. Denison, 1979. "Empirical Atmospheric Deposition Parameters - A Survey," Atmospheric Environment, 13, p. 1000.

Praham, L.P. and Berkowicz, R., 1978. "Predicting Concentrations in Plumes Subject to Dry Deposition," Nature, 271, p. 232-234.

Rao, K.S., 1986. User's Guide for PEM-2 - Pollution Episodic Model (Version 2). Office of Research and Development, U. S. Environmental Protection Agency.

Sehmel, G.A., 1980. "Particle and Dry Gas Deposition: A Review," Atmospheric Environment 14, 1002.

Sehmel, G.A., and W.H. Hodgson, 1974. "Predicted Dry Deposition Velocities" in Atmospheric-Surface Exchange of Particulate and Gaseous Pollutants. Technical Information Center, Energy Research and Development Administration, U. S. Department of Energy.

TACB, 1979. User's Guide to the Texas Episodic Model, TACB, Permits Section, Austin, TX.

Yamartino, R.J., 1981. "Atmospheric Pollutant Deposition Modeling," in Handbook of Meteorology, McGraw Hill, New York, N.Y.

ACCOUNTING FOR WET DEPOSITION IN INCINERATOR RISK ASSESSMENTS

Sally A. Campbell[1], Kenneth L. Zankel[2], Roger Brower[2], and James M. Teitt[3]

[1]S.A. Campbell Associates, 10714 Midsummer Lane, Columbia, MD 21044, [2]Versar, Inc., ESM Operations, 9200 Rumsey Road, Columbia, MD 21045, 3Power Plant & Environmental Review Division, Tidewater Administration, Maryland Department of Natural Resources, Tawes State Office Building, Annapolis, MD 21401

INTRODUCTION

The deposition of materials from the atmosphere is a critical link in the pathway by which toxic atmospheric pollutants (TAPs) are transported to crops and other surface materials. This deposition occurs by dry processes with some efficiency, particularly for gases and large particles (greater than 1 micron). However, in the presence of rain and other hydrometeors, deposition of many TAPs is greatly enhanced. This enhancement occurs both because the presence of rain and snow greatly increases the active surface area presented to atmospheric contaminants and because the hydrometeors tend to sweep out a volume in their fall, collecting particles that are not displaced by the passing hydrometeor. In addition, hydrometeors falling through plumes close to discrete sources are exposed to relatively high concentrations of toxic materials at the plume centerline that may never be experienced at ground-level.

When locally grown food is a major dietary component, experience has shown that contamination of the food chain through wet and dry deposition of non-volatile TAPs on particulate material dominates risk estimations for combustion sources (PPER 1989). Of the two deposition routes, wet deposition is by far the more significant for estimating maximal risks, potentially driving both the location of the maximally exposed individual (MEI) and the MEI risk magnitude for those TAPs associated with particles.

The importance of particle wet deposition relative to dry deposition decreases with distance (as discussed below). For the average exposed individual in farming areas (AEI), wet deposition may be expected, on theoretical grounds, to produce risks of the same order of magnitude as those attributable to

dry deposition and significantly greater than the direct inhalation risk. Only for urban settings where food chain exposures are slight is neglect of either of these two ingestion
exposure routes warranted.

The term wet deposition comprises a group of processes
that share only the feature that they involve condensed water
vapor. This group includes:

o   Rain out: generally applied to deposition of
      materials captured by droplets within clouds either
      during or after droplet formation.

o   Wash out: generally applied to scavenging by rain
      below the cloud base.

o   Scavenging by snow, sleet and hail.

o   Deposition of fog laden with materials.

o   Scavenging during dew formation.

In addition, washdown, the transport by rain of materials dry
deposited to natural and man-made surfaces, must be considered
in any attempt to estimate total environmental risk for a discrete source. For instance, materials deposited to crops,
buildings, and parking lots will eventually be transported to
soils and surface waters.

Of these routes, only rain out, wash out and washdown are
commonly considered in risk assessments at the present time.
Night-time dry deposition rates to wet surfaces are sometimes
computed, but the more commonly used treatments (e.g., Shieh
et al. 1979; Hicks et al. 1987) do not incorporate information
on the diffusiophoretic effects that occur during dew formation.

In this report, we review the general characteristics and
significance of wet deposition of TAPs, briefly describe the
major approaches currently used to assess it, and present some
detailed information on features of the computations that are
particularly troublesome. Finally, we summarize the information that is still needed to put the present approaches on a
firm scientific basis. Our emphasis in this work is on the
deposition of TAPs by rain and precipitating clouds, as this
is the most well developed element of the estimation. In
keeping with the theme of this conference, we have focused on
estimating deposition in the vicinity of discrete sources.

APPROACHES TO ESTIMATING WET DEPOSITION

The general mechanism of wet deposition rather simply
involves the uptake of particles and gases in hydrometeors and
their subsequent transfer to the ground. However, estimating
wet deposition involves a great many steps, many of which are
not well defined. These may be grouped as follows:

o   Defining spatial and temporal distributions of
      atmospheric concentrations for the species of
      interest.

o   Defining spatial and temporal distributions of rain
     or snowfall or the occurrence of fog and dew.

o   Estimating the collection efficiency of the
     hydrometeor for the species of interest.

o   Estimating changes in atmospheric concentrations for
     the species of interest as deposition proceeds.

The problem cannot be readily simplified by using long
term averages because atmospheric concentrations in the vicinity of discrete sources tend to be highly correlated (covari-
ant) with meteorological conditions favoring precipitation
events.

Estimating Deposition Due to Rain and Frozen Precipitation

Two major approaches are presently used for estimating
deposition due to rain and frozen hydrometeors (snow, sleet
and hail). The first method addresses in detail the micrometeorological events that lead to deposition. This approach,
embodied in Mesopuff (EPA) and other major models, assumes
that wet deposition is proportional to the atmospheric concentration, χ, and that the constant of proportionality may be
thought of as a scavenging coefficient, λ. This coefficient
is a function of scavenging efficiency, precipitation rate,
and species-specific parameters such as size distributions of
particles and solubility of gases. That is, the wet deposition flux is given by

$$Dep_w = \int_{0}^{z_w} \lambda \chi \, dz$$

where Z_w is the vertical extent of scavenging by the precipitation (Bowman et al. 1987).

Within cloud and below cloud events are frequently distinguished. The necessary scavenging coefficients may be
measured or estimated from theoretical considerations. Bowman
et al. (1987) has recommended a set of particle scavenging
coefficients for use in wet deposition modeling using a standard Gaussian dispersion model. However, various alternative
values have been suggested in recent years. Currently used
approaches for estimating for gases and particles are
described in Sections III and IV (below), respectively.

In the second major approach, a washout ratio, w_r, is
defined as the measured average ratio of the concentration in
rain or snow to the (contemporaneous) concentration in air at
a specific location. These ratios are commonly expressed as
the concentration ratio, $[(g/l_{water})/(g/l_{air})]$, although values
are sometimes reported in alternative units. Measured values
of washout ratios for particles and gases range from 0 to 10^6
and tend to be in the range of 10^3 to 10^5 for submicron particles (Slinn 1984; Ligocki 1985b).

The use of washout ratios in estimating deposition avoids
the need for complicated modeling computations. Furthermore,

for any given case, they automatically account for effects that are difficult to model a priori such as particle growth; enhancement of gas solubilities due to chemical reaction; and pollutant redistributions within the atmosphere. For this reason, washout ratios may be applied under very limited conditions for simple problems such as estimating the effect of simple emissions changes at a single source or location. They may also provide a useful "sanity check" (order of magnitude agreement) on more theoretical treatments applied to the conditions under which the measurement is made.

However, washout ratios suffer from a number of drawbacks; most notably, these ratios are specific to the site at which they were determined. They do not include the information necessary to correct for variations in plume depth, in rainfall rate and amount, or in the mix of emitted reactive contaminants. These variations are known to be important (Slinn, 1984), and may dominate the computation of deposition rates when properly accounted for. Thus, washout ratios cannot be generalized to provide short term deposition estimates, or estimates for locations, sources, and meteorological conditions dissimilar from the ones where the measurements were made.

Hosker (1980) has suggested a method for computing the scavenging coefficient from washout ratios:

$$\lambda = \frac{W_r\, R}{Z_w}$$

where R is the rainfall rate.

However, this conversion must be used with great care: not only does it perpetuate the difficulties with washout ratios described above, but it introduces new assumptions about the nature of the contaminated atmosphere which have to be tested. For instance, if the washout data have been obtained for plumes, they can only be used for similar emissions sources (because size distributions must match). If they have been obtained from observations in the free atmosphere, not only size distributions but plume depths must be inferred. Finally, for slightly soluble, non-reactive gases, washout ratios and scavenging coefficients computed from them have little applicability (see Section III below).

Approaches for estimating deposition by snow, sleet and hail are similar to those used for rain. Slinn (1984) presents recommended data derived from theoretical considerations for computing scavenging coefficients and for their incorporation into models. These recommended coefficients would result in somewhat lower estimations of deposition by snow. However, deposition by snow is even more complicated than rain due to the large variety of different snow forms. Measurement data is meager and there is little agreement between published values. For example, data for scavenging by snow of 0.5 micron particles show variations of about 2 orders of magnitude. However, the few available measurements of snow scavenging coefficients do not support the notion that snow anticipated to affect exposures to elevated emissions close to the source: in fact, the presence of fogs tends to strengthen ground-based inversions and may actually reduce the local incidence of direct plume impaction from that which would be estimated from routine plume modeling routines. Rather, fog production (except deep layer fogs) would produce increased exposure and deposition only at distant locations where the plume has become mixed to the ground or possibly in land-sea circulation cells that return elevated emissions to the land at low levels.

Direct deposition from non-precipitating clouds may be particularly important at high elevations. Such clouds are known to provide significant water inputs on mountain tops (Falconer and Falconer 1979; Kyriazopoulos et al. 1978). Water collection volumes may exceed deposition due to rain at these locations and may be highly enriched with TAPs scavenged from the atmosphere at high elevations. In cases where the combustion plume encounters clouds at the height of the terrain, this mechanism may significantly increase deposition (and runoff to water bodies) of the TAPs. To our knowledge, TAP concentrations have not been determined in mountain top cloud water collections, and fog and cloud effects have not been taken into account in existing risk assessments for discrete sources.

For risk assessment purposes, most of the necessary methods are available to estimate deposition of water droplets from fog or non-precipitating clouds to plant canopies, and also to estimate enhanced lung deposition of toxic materials from fogs. However, no information was found on the occurrence of the more significant TAPs in fog and cloud. A number of experiments were conducted in the early part of this decade to measure the concentration and deposition of materials from ground fog and from clouds impacting mountaintops (Dollard and Unsworth 1983; Brewer et al. 1983; Waldman et al. 1982; Black and Landsburg 1983; Falconer and Falconer 1979). However, most of these experiments have addressed only those ions related to acidic deposition.

Presumably, the methods being developed for estimating with-in cloud scavenging of gases and particles could be applied to estimate the composition of fog and cloud water, but these computations cannot be used with confidence until experimental data have been developed to test the theoretical treatments.

Ground fogs differ from rain and precipitating clouds in that significant chemical reactions between components may occur during their relatively long exposure to fresh ground-level emissions, and these must be taken into account when computing their composition. In particular, uptake and modification of toxic organic compounds may proceed efficiently when the fog forms in an urban atmosphere rich in nitrating and oxidizing gases and metallic catalysts. These reactions may convert relatively non-toxic materials to more toxic forms and, conversely, detoxify other materials. For instance, some research indicates that both oxidation and nitration of polycyclic aromatic hydrocarbons, PAHs, proceeds rapidly in water droplets and presumably in contaminated fogs (see reviews by Finlayson-Pitts and Pitts 1986 and Campbell 1989). Reaction

Table 1. Comparison of data for scavenging of particles by
          rain and snow

Scavenging Coefficient	Precipitation Type	Source
0.5×10^{-5}	snow	Wolf and Dana (1969)[a]
$18 - 28 \times 10^{-5}$ (1977)[a]	snow	Gradel and Franey
5×10^{-5}	snow	Engleman et al. (1966)[b]
20×10^{-5}	rain	Bowman et al. (1987)
4×10^{-5}	rain	Schumann et al. (1988)

[a]Cited in McMahon and Dennison (1979)
[b]Cited in Slinn (1984)

scavenging of small particles is less efficient than scavenging by rain (Table 1).

Estimating Deposition Due to Fog and Cloud Water

     Radiative ground fog formation may be expected to in-
crease the risk due to pollutants in the atmosphere both by
providing a new deposition route and by collecting submicron
particles that may not deposit efficiently in the lung and on
vegetation, presenting them in a particle size that is well
collected. Waldman et al. (1982) have demonstrated that concentrations of both soluble gases and insoluble material that
is usually associated with particles is much higher in fog
water than in the free atmosphere. Dollard and Unsworth
(1983) demonstrated that deposition of wind blown fog to
grassland was on the order of 0.14 mm/8 hours and could be
estimated using methods commonly used for addressing dry
deposition rates (e.g., multi-layer linear resistance dry
deposition models). By analogy with other dry deposition
findings, fog deposition to trees and other dissected vegetation may be expected to be on the order of 0.5 to 1 mm/8
hours. Because of the high concentrations of materials anticipated to occur in fog, the annual deposition of toxic materials by this route may be comparable in magnitude at some
locations to rainwater input.

     In addition, soluble submicron particles exhibit rapid
increases in size at moderate to high relative humidities, and
this size increase greatly enhances their "dry" deposition to
surficial features and during inhalation. Therefore, the high
humidity associated with fog formation may enhance both the
food chain and the inhalation risks associated with emissions.

     However, because ground fogs occur only under conditions
of extreme ground-level stabilities, they would not be

anticipated to affect exposures to elevated emissions close to the source: in fact, the presence of fogs tends to strengthen ground-based inversions and may actually reduce the local incidence of direct plume impaction from that which would be estimated from routine plume modeling routines. Rather, fog production (except deep layer fogs) would produce increased exposure and deposition only at distant locations where the plume has become mixed to the ground or possibly in land-sea circulation cells that return elevated emissions to the land at low levels.

Direct deposition from non-precipitating clouds may be particularly important at high elevations. Such clouds are known to provide significant water inputs on mountain tops (Falconer and Falconer 1979; Kyriazopoulos et al. 1978). Water collection volumes may exceed deposition due to rain at these locations and may be highly enriched with TAPs scavenged from the atmosphere at high elevations. In cases where the combustion plume encounters clouds at the height of the terrain, this mechanism may significantly increase deposition (and runoff to water bodies) of the TAPs. To our knowledge, TAP concentrations have not been determined in mountain top cloud water collections, and fog and cloud effects have not been taken into account in existing risk assessments for discrete sources.

For risk assessment purposes, most of the necessary methods are available to estimate deposition of water droplets from fog or non-precipitating clouds to plant canopies, and also to estimate enhanced lung deposition of toxic materials from fogs. However, no information was found on the occurrence of the more significant TAPs in fog and cloud. A number of experiments were conducted in the early part of this decade to measure the concentration and deposition of materials from ground fog and from clouds impacting mountaintops (Dollard and Unsworth 1983; Brewer et al. 1983; Waldman et al. 1982; Black and Landsburg 1983; Falconer and Falconer 1979). However, most of these experiments have addressed only those ions related to acidic deposition.

Presumably, the methods being developed for estimating with-in cloud scavenging of gases and particles could be applied to estimate the composition of fog and cloud water, but these computations cannot be used with confidence until experimental data have been developed to test the theoretical treatments.

Ground fogs differ from rain and precipitating clouds in that significant chemical reactions between components may occur during their relatively long exposure to fresh ground-level emissions, and these must be taken into account when computing their composition. In particular, uptake and modification of toxic organic compounds may proceed efficiently when the fog forms in an urban atmosphere rich in nitrating and oxidizing gases and metallic catalysts. These reactions may convert relatively non-toxic materials to more toxic forms and, conversely, detoxify other materials. For instance, some research indicates that both oxidation and nitration of polycyclic aromatic hydrocarbons, PAHs, proceeds rapidly in water droplets and presumably in contaminated fogs (see reviews by Finlayson-Pitts and Pitts 1986 and Campbell 1989). Reaction limits based on considerations such as product solubility can not be used to estimate the extent of the reactions since the products may well be gaseous components that are released from the fog to the free atmosphere. This situation is still very confused and needs to be evaluated by measurements under environmentally relevant conditions.

The computation of the contribution of fog formation to risk is further complicated by the difficulty of estimating the occurrence and duration of local fog events from commonly available meteorological data (on-site observations are commonly lacking for the areas of greatest interest). The evolution of fog droplet size and composition with time must also be taken into account in estimating total deposition both to surfaces and during inhalation (see for instance, Waldman et al. 1982).

DEPOSITION OF GASES (AND VAPORS OF VOLATILE COMPOUNDS) IN RAIN

The mechanisms governing the scavenging and deposition of gases and volatile compounds are fundamentally different from those governing the deposition of particles. In particular, scavenging of particles is irreversible, and particles collected by rain in passing through a plume are inevitably deposited at ground-level (Fig. 1a). By contrast, the scavenging of volatile compounds is generally reversible, and rainwater concentrations at ground-level may bear little relationship to atmospheric concentrations aloft (Fig. 1b). Although the general concepts of scavenging coefficient and exponential decay have been applied in estimating deposition of gaseous components, they apply in reality only to conditions in which the concentration in the boundary layer shows little vertical variability. This section describes the approaches presently used in estimating wet deposition of gases from elevated plumes.

Scavenging of Volatile Compounds

The deposition of gases and volatile compounds in rain depends on their dissolution in water. This dissolution is generally reversible, and thus rain falling through an elevated plume may release pollutants as it traverses the relatively clean air below the plume. The rate at which equilibrium is achieved depends both on the diffusion of contaminants through the atmosphere to the droplet and on the diffusion of contaminants through the droplet interior. Under most conditions, diffusion within the water droplet is rate-limiting. For very soluble substances, achievement of equilibrium may be slow, and rain concentrations are related to elevated plume centerline concentrations. For such vapors and gases (dimensionless Henry's Law constants greater than 5×10^{-3}) raindrop concentrations equilibrate with air concentrations within about 10 meters. For these components then, the equilibrium concentration in the droplet is determined almost entirely by the ground-level concentration (GLC) of the pollutant. Furthermore, the location of maximal wet deposition will occur at the location of highest ground-level concentration.

34

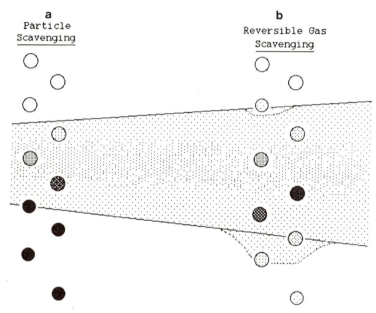

Fig. 1.   Scavenging of particles (1a) and gases (1b) by
           rainwater

Since 1) dry deposition is also dependent on the ground-level concentration and 2) maximal dry deposition will similarly occur at the location of highest GLC, it is instructive to compare the maximal wet and dry deposition rates for slightly soluble gases. Dry deposition of gases may be described by the well known relationship:

$$Deposition = deposition\ velocity \times concentration$$

For gases, measured deposition velocities (V_d) are commonly in the range of 0.01 cm/sec to 1.5 cm/sec for vegetation (McMahon and Denison 1979; Campbell 1989) and the annual total deposition (in g/cm^2) may be approximated (ignoring the covariances mentioned in Section II for the moment) as 3×10^7 sec/yr $\times V_d \times C$, where C is the mean ground-level atmospheric concentration.

For wet deposition of slightly soluble gases, the mass deposition per unit area is given by

$$Deposition = V \times C \times H$$

where V is the annual rainfall depth, C is as defined as above, and H is the dimensionless Henry's Law constant (HLC). Annual wet and dry deposition rates are approximately equal for the condition

$$V \times H = 3 \times 10^7 \times V_d$$

Thus, for a typical rainfall amount of 1 m/year and dry deposition velocity of 0.1 cm/sec, dry deposition exceeds wet deposition for values of H less than about 3×10^4. In general, then, wet deposition may be safely neglected for most slightly soluble organic vapors. To allow for the effect of covariances and to ensure no more than a 10% error in deposition, wet deposition should be calculated whenever the ratio of H (dimensionless) to annual average deposition velocity is suspected to be greater than about 3×10^5 sec/m.

Even though wet deposition of slightly soluble gases may not be significant by comparison with dry deposition, the reversible absorption of pollutants by rain will cause a "tilting" effect as pollutants are absorbed at plume centerline and released below the plume. This effect would tend to increase the ground-level concentration and inhalation risk at some locations. To our knowledge, the magnitude of this plume tilting has not been evaluated for conditions relevant to incinerator plumes.

Selection of Henry's Law constants for slightly soluble materials is very important. Since direct measurements are rather difficult to make and not available for many constituents (because of their extremely low equilibrium concentrations in water), many investigators have used the ratio of the water solubility to the vapor pressure of the constituent. HLC values derived this way are theoretically correct only for constituents that form "ideal" solutions. In fact, many of the TAPs of interest from incinerators do not obey ideal solution theory. For these components, it is important to find

and employ HLC values obtained by direct measurement or
obtained from more sophisticated calculation procedures.

A variety of HLC values have been developed for the less
soluble TAPs using activity coefficients and solubilities calculated by the UNIFAC method. Recently, this method was found
to contain systematic errors leading to significant errors in
solubilities and related HLC estimations. Banerjee and Howard
(1988) describe a correction factor to improve agreement
between results obtained using UNIFAC and available measurements; even these corrected estimates differ by as much as an
order of magnitude from measured values. This issue is currently the subject of a lively discussion in the literature
that may be anticipated to provide greatly improved estimates
of activity coefficients, water solubilities, octanol-water
distribution coefficients, and related properties for very
slightly soluble compounds. Even more important, this discussion has added impetus to attempts to develop direct
methods for measuring these quantities.

Because computational and measurement methods are rapidly
evolving, care must be taken in selecting HLCs for use in
deposition modeling to ensure that they represent the latest
available measurement or estimation methods. Finally, correction of HLCs determined for 25°C to the actual temperature of
the rain and ambient atmosphere is required: Ligocki et al.
(1985a) report that this correction may change the deposition
estimate by a factor of 3 to 6 when rain water temperature is
in the range of 5 to 10°C.

The Semivolatile Problem

In the preceding section, we have indicated that wet
deposition may often be neglected for slightly soluble gases.
On the other hand, wet deposition of particulate materials is
often more significant for risk assessments than is dry deposition. Many of the TAPs commonly considered in incinerator
risk assessments are semivolatile, that is they are reversibly
released from atmospheric particles. The vapor fraction is,
of course, a function of temperature.

For these compounds, multiple pathways must be evaluated
in estimating their deposition. On the one hand, the vapor
and particulate fractions can each be expected to deposit in
the usual manner. In addition, the uptake and loss of the
compound to and from scavenged particles in rain drops and at
the ground must be considered (Fig. 2). In addition, the
effect of the water environment of the drop must be considered
in estimating exchanges between the various compartments.
Pankow (1987), Ligocki et al. (1985 a,b) and Bidleman (1988)
may be consulted for information on the scavenging of vapors
by atmospheric particles and of selected semivolatile
constituents by rain.

Neither the theoretical nor the experimental base is in
place to allow proper evaluation of the deposition of many
semi-volatile TAPs. As an interim measure, we suggest that
the four deposition routes be evaluated (separate dry and wet
deposition routes assuming compound is entirely a gas or
entirely particulates) with the highest resulting value being
selected for risk computations.

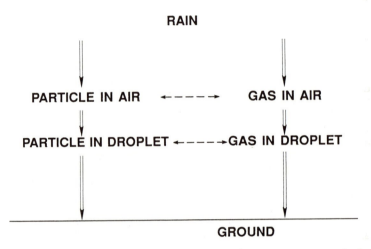

Fig. 2.   Routes of mass exchange during wet deposition of a
semivolatile compound

WET DEPOSITION OF PARTICLES

Because rain drops are very efficient at collecting particles and bringing them to the ground, wet deposition will
usually dominate risk at the MEI in the vicinity of a point
source, such as a municipal incinerator. It is, therefore,
extremely important to have reliable methods available to
project wet deposition of pollutants emitted from point
sources. Although considerable research has been performed on
methods to estimate scavenging of particulate material by
rain, current methods can lead to very large (order of magnitude) uncertainties when projecting point source impacts at
the MEI.

Neither theory nor experimental data are sufficient to
allow accurate estimations of particle scavenging coefficients. Although very sophisticated theoretical models have
been developed for the scavenging of particulate material by
precipitation, they only explain scavenging in a qualitative
manner. Due to uncertainties in the values for the input
parameters needed to exercise those models, it is not feasible, at this time, to rely on theoretical models for projecting close-in deposition due to emissions from an
incinerator. Furthermore, the models are highly complex and
require highly intensive computer resources, even for this day
and age. Very few experimental studies that have been performed are applicable to close-in impacts due to deposition of
particles originating from a point source. Conditions that
determine deposition are highly variable, making it impractical to rely solely on the use of experimental results.
Thus, current methods of projecting impact employ a semi-
empirical approach, which relies on the theoretical principles
and the few available experimental results.

In the following, we address the theoretical concepts
related to close-in wet deposition of particles and the models
currently used for estimating wet deposition.

Theoretical Concepts

We will make no attempt here to completely review the
theories on scavenging. For this, the reader is directed to
reviews by Slinn (1984) and Prodi and Tampieri (1982).
Instead we focus on those aspects that we feel important to an
understanding of the difficulties encountered and on those
concepts that may be useful for interpreting semi-empirical
evaluations of wet deposition.

The removal of a pollutant from the atmosphere by a hydrometeor (rain drop or snow flake) is governed by the impaction of aerosol particles with a hydrometeor caused by the
relative motion of the two (including Brownian motion of the
particle). On the one hand, small particles are likely to
avoid impaction by going around the particles rather than
impacting directly. On the other hand, very small particles
are subject to increased motion (Brownian motion), which makes
it more likely for impaction to occur. This leads to a scavenging gap for particles with diameters in the range of about
0.1 to 1 micron. Particles very much larger than this will
not be able to avoid the hydrometeors and will impact

efficiently. Particles very much smaller will be collected
efficiently due to their Brownian motion.

There is much uncertainty regarding the location and
extent of the scavenging gap. Figures 3a, 3b and 3c taken
from Radke et al.(1980) present data that illustrate the
scavenging gap. (Efficiencies in the figures are relative to
direct sweeping out of the particles by gravitational settling
of rain, ignoring Brownian motion or avoidance of the rain
drops by the particles.) Note that the experimental results
depicted in the figures are highly variable with regard to the
location of the gap and the efficiencies of collection. Theoretical results, shown in Fig. 3d, also indicate a scavenging
gap (the increase due to Brownian motion for some of these
cases is too far to the left of the figure to be displayed).
However the location, width, and extent of the gap differ
considerably from the experimentally derived results. Differences between theory and experiment are probably due to an
inability to properly account for the many effects that govern
motion such as electrophoresis and micrometeorological gradients. Nucleation of hydrometeors by particles in clouds,
aggregation of particles, and deliquescent growth of particles
further complicate the interpretation of experimental results.
It is assumed in the theory that once a particle impacts with
a hydrometeor, it sticks. There is a possibility that the
sticking coefficient would depend on the type of particle.
Furthermore, for a given amount of rain, the area swept out by
the raindrops is highly dependant on the size distributions of
the raindrops; and the size distribution of the raindrops
varies considerably. Since the location of the gap varies,
even within a rain event (see Fig. 3a), the width of the gap
will become much wider and the depth much less pronounced when
averaged over a number of rain events.

The uncertainties associated with the scavenging gap
become particularly important for determining impacts from
municipal incinerators (as well as for other controlled combustion sources) because a large portion of the pollutant mass
is in the size range at the gap. Table 2, taken from Weston
(1988), shows the measured size distribution of particles
emitted from an incinerator in Munich. If one assumes that
certain pollutants, such as dioxins, are distributed on the
surface, rather than throughout the particles, the distribution shifts towards smaller sizes. Most of the mass is then
distributed in the vicinity of the scavenging gap. Clearly it
is important to know where, and to what extent, the gap occurs
and the detailed size distribution of the pollutant, particularly in the vicinity of the gap. These problems become exa-
cerbated when attempting to determine deposition of particles
emitted from stacks using only size distribution data taken at
the stack. Particles can become considerably larger due to
deliquescence and/or aggregation, which could shift the distribution from primarily within the range of the scavenging
gap to without.

Although it is not possible, at this time, to accurately
determine deposition from purely theoretical calculations, the
theory furnishes a number of insights valuable for interpreting measurements and designing additional experimental
studies. Clearly, deposition will depend on the size distribution of raindrops. This varies considerably with the type

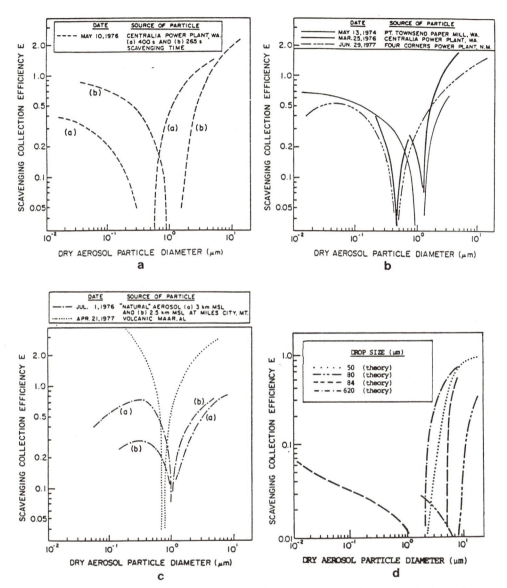

Fig. 3.   Measured and theoretical scavenging efficiencies
(taken from Radke et al. (1980)).

Table 2. Size distributions for the Munich incinerator

Diameter (Microns)	Mass (%)	Area (%)
0.09	9.1	60.4
0.4	10.1	15.1
0.7	16.2	13.8
1.1	10.1	5.5
2.3	8.1	2.1
3.7	6.0	1.0
5.4	4.8	0.5
8.0	7.4	0.6
12.9	9.1	0.4
18.0	19.1	0.6

of rain event. Frontal storms, for example, would furnish different results than convective storms. The movement of particles by other than Brownian motion could also vary considerably with meteorological conditions. This movement also includes motion due to electrophoretic and thermophoretic forces. Because of the high variability in the parameters affecting scavenging, measurements based on limited types of rainfall could not be used generally.

For plumes that are below rain clouds (for wash-out), the amount of pollutant intercepted by the rainfall is, to the first approximation, independent of the vertical dispersion of the plume, i.e., the probability that a pollutant molecule is scavenged by rain is independent of the height of the molecule above ground. However, as the plume travels away from the source, it disperses horizontally. Rain falling through the center of the plume will encounter fewer pollutant molecules at larger distances from the source. One then would reasonably expect the maximum impact due to wet deposition to occur at the facility boundaries. Further lateral dispersion occurs when averaging over a large number of rain events, with different wind directions. As a result, we should expect the average deposition to vary inversely with distance from the source.

If the plume enters the clouds, removal of the particulate material by hydrometeors (leading to rain-out) proceeds more rapidly. However, for impacts to the MEI, we are concerned not by the removal of particles from the plume, but by particle containing rain falling to the ground. While the deposition of pollutants by raindrops falling through the plume occurs almost directly beneath the location of impaction, the deposition of pollutants removed by a cloud can occur many kilometers downwind (see, for example, Gatz (1977)).

Current Models and Comparisons With Field Studies

Currently used models for determining wet deposition assume that the plume is below the clouds, that the deposition is proportional to the vertically integrated concentration of the plume, and that, as deposition continues, the plume mass is reduced by $exp(\lambda t)$. The amount deposited is therefore the vertically integrated concentration multiplied by λt. The scavenging coefficient, λ, is obtained from previous measurements on scavenging. Scavenging coefficients recommended by Bowman et al. (1987) and adopted by EPA (1986) are shown in Table 3. Different scavenging coefficients are presented for different rates of rainfall, in categories usually recorded by the National Weather Service. Scavenging coefficients are also given for three different particle size ranges. No discrimination is made for rainfall type.

The coefficients recommended by Bowman were based on results of Radke et al. (1980). Scavenging was calculated from the number size distribution of particulates before and after being subjected to a rain event. Some of the cases studied involved scavenging of particulates that were generally dispersed in the atmosphere, others involved scavenging of plumes. Most of the data was apparently collected when in-cloud scavenging predominated (Moore and Anderson 1988). A list of the sources studied and the meteorology during the studies are shown in Table 4. Because in-cloud scavenging is more efficient than below-cloud, the recommended coefficients probably lead to overestimates of deposition from plumes beneath the clouds. Furthermore, since scavenging rather than deposition was measured, maximum deposition would be overestimated for in-cloud scavenging events unless transport in the clouds before rain-out is considered. This can not be done with a simple exponential decay model.

Employing similar measurement techniques to those of Radke cited above, Schumann et al. (1988) recently measured below-cloud scavenging of particulates by rainfall. The events studied are listed in Table 5. Three of these events furnished data sufficient for use in determining scavenging coefficients as a function of particle size. The averaged results (solid curve) of these three cases is shown in Fig. 4. The area bounded by the dashed curves indicates the considerable amount of scatter between the three cases. It can be seen that the scavenging coefficient increases rapidly for particles larger than a few microns. For this limited number of cases, the rainfall rates were in the range of light precipitation as defined for the Bowman coefficients. We used the results of the three applicable cases from Schumann and calculated scavenging coefficients for the three size intervals used by Bowman. For each particle size interval, scavenging was calculated as an average, weighted by the size distribution for a pollutant on the surface of particulates emitted from the Munich municipal incinerator (Table 2). Table 6 illustrates that, as expected, the scavenging coefficients obtained from below-cloud data are lower than those obtained from within cloud data.

Table 3.   Scavenging coefficients (s^{-1}) as a function of
           particle size and precipitation intensity

Particle Size

Precipitation intensity[a]	< 2 μm	2 μm–10 μm	> 10 μm
Light	$2.20\text{x}10^{-4}$	$1.8\text{x}10^{-4}$	$9.69\text{x}10^{-3}$[b]
Moderate	$5.60\text{x}10^{-4}$	$8.93\text{x}10^{-4}$	$9.69\text{x}10^{-3}$[b]
Heavy	$1.46\text{x}10^{-3}$	$4.64\text{x}10^{-3}$	$9.69\text{x}10^{-3}$[b]

[a] Light -- less than 0.1 inches per hour (< 2.5 mm/hr)
    Moderate -- 0.11 to 0.30 inches per hour
              (2.5 mm/hr-7.6 mm/hr)
    Heavy -- greater than or equal to 0.31 inches per hour
           ($\geq$ 7.7 mm/hr).
[b] Bowman et al. (1987) indicate that insufficient data were
    available for the largest particle size to perform a
    meaningful least-square fit to produce scavenging coeffi-
    cients as a function of precipitation intensity. The
    average value given by Bowman et al. is used here.

        We are faced with a dilemma. On the one hand we have data
for in-cloud scavenging that is probably not applicable to
below cloud scavenging. On the other hand, there is limited
data for below cloud scavenging, and only for light rainfall
rates. Scavenging coefficients for higher rainfall rates
could be estimated by assuming that the scavenging coefficient
is proportional to $R^{0.75}$ as suggested by Slinn (1984). The
results of performing this calculation are shown in Table 7.
For plumes that get entrained by clouds, the situation becomes
even more complex. There is increased scavenging that is off-
set by the added dispersion due to travel before the particles
come down in rainfall. Clearly the technique of incorporating
a scavenging coefficient into existing point source dispersion
models is inadequate to describe deposition from in-cloud
plumes. Use of such a dispersion model with the Bowman coefficients would be extremely conservative. For convective
storms, the usual dispersion models, such as ISC, would be
inadequate to even determine whether large portions of the
plume are convected into the clouds. Furthermore, for plumes
entering clouds, the location of maximum impact would be far
different in reality from what would be predicted from such a
model. It is possible that the added dispersion (including
that due to the longer travel time) makes the long-term
average contributions of in-cloud incinerator plumes to the
impacts on the MEI negligible compared to the contributions
from below-cloud plumes. Further study is needed on this
matter.

Table 4. Conditions under which measurements were obtained (taken from Radke et al. (1980))

Date	Source of Scavenged Aerosol Particles	Location	Nature of Cloud and Precipitation Scavengers	Scavenging Time (s)	Average Rain Rate (mm h^{-1})
5/13/74	Kraft-process paper mill	Port Townsend, Washington	Cumulonimbus with low cloud base; rain shower	960	7
3/25/76	Natural	Near Centralia, Washington	Cumulonimbus with medium cloud base; rain shower	480	18
5/10/76	Coal-fired power plant	Near Centralia, Washington	Stratocumulus precipitation with an orographic component; rain	(a) 400 (b) 265	10
7/1/76	Natural	Near Miles City, Montana	Cumulonimbus cluster, high cloud base, lightning discharges; heavy rain showers	(a) 1800 (@ 3 km MSL) (b) 1100 (@ 2.5 km MSL)	9
4/21/77	Emissions from a volcanic maar	Near King Salmon, Alaska	Cumulonimbus with low cloud base; graupel shower	300	12
6/29/77	Coal-fired power plant (2100 MW)	Near Farmington, New Mexico	Isolated cumulonimbus with high cloud base; short period of light rain showers	600	8

Table 5. Observed precipitation events and their characteristics studied by Schumann et al. (1988)

Date	Synoptic Situation	Precipitation		
		Total (mm)	Maximum Duration (h)	Rate (mm h^{-1})
1/10/86	Warm front with trapped cold sector	4.2	11.5	1.4
1/22/86	Weak cold front in south westerly flow	0.3	4	1.2
3/19/86	Weak surface low producing stagnation (cold)	6.7	4.5*	4.8

*End of observation (precipitation not ceased).

The available data on deposition due to snow is even less than for rain. Because of the larger surface area for the same amount of precipitation expressed as water, it is possible that snow is a more efficient scavenger than rain. Based on the limited information available, Bowman (1987) suggests using the same coefficients for snow as he proposed for rain. Clearly, more measurements are needed.

CONCLUSIONS

The procedures for estimating wet deposition of TAPs from discrete source plumes are in varying degrees of disarray. Although the general methods for particles and volatile gases are well established, many of the critical input data are either poorly known or little understood.

For particle deposition, it would be desirable to have a formalism that relates deposition to readily predicted parameters such as typical size distributions of the emitted pollutants and readily available meteorological data. Size distributions are presently not known for the individual pollutants, changes in size distribution due to aggregation and deliquescence of the particles complicates matters, and the dependance of scavenging on a whole host of meteorological variables makes determining average scavenging coefficients difficult to obtain. Currently, scavenging coefficients are subdivided by size and rainfall rate categories. In order to project long-term average impacts reliably, further data is needed to determine best values for the parameters, to determine if and what further subdivisions are needed to account for variability in rainfall type, and to determine whether more size categories are needed to account for the scavenging gap. Methods must be developed for handling deposition due to in-cloud scavenging of the plume.

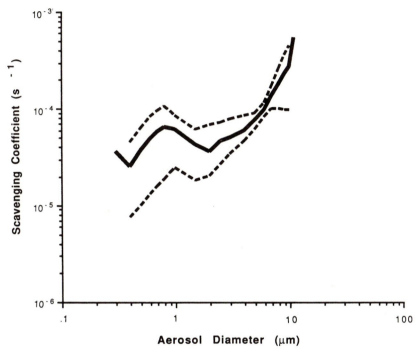

Fig. 4.   Average and range of scavenging coefficients as a
          function of particle size for three cases found in
          Schumann et al. (1988)

Table 6. Scavenging coefficients (s^{-1}) as a function of
         particle size for light precipitation intensity[a]

		Particle Size	
Source of data	< 2 μm	2 μm–10 μm	> 10 μm
Radke et al. (1980)	2.20×10^{-4}	1.80×10^{-4}	9.69×10^{-3}[b]
Schumann et al. (1988)	4×10^{-5}	6×10^{-5}	6×10^{-4}

[a] Light -- less than 0.1 inches per hour (< 2.5 mm/hr).
[b] Bowman et al. (1987) indicate that insufficient data were
   available for the largest particle size to perform a
   meaningful least-square fit to produce scavenging efficients
   as a function of precipitation intensity. The average value
   given by Bowman et al. is used here.

Table 7. Scavenging coefficients (s^{-1}) based on Schumann et al.
         (1988) data

	Particle Size		
Precipitation Intensity	< 2 μm	2 μm–10 μm	10 μm–20 μm
Light	4×10^{-5}	6×10^{-5}	6×10^{-4}
Moderate	1.1×10^{-4}	1.7×10^{-4}	1.7×10^{-3}
Heavy	2.2×10^{-4}	3.4×10^{-4}	3.4×10^{-3}

For deposition of slightly soluble gases the major unknowns are in the area of estimating Henry's Law constants. It is hoped the present intensive efforts to improve and validate the methods for estimating these data from theoretical calculations will soon yield reliable values that can be used with confidence in computing deposition rates. In particular, improvements in techniques for direct determinations of HLCs will provide valuable information for validating theoretical routes to this information.

For deposition of semivolatiles, development and validation of theoretical (non-parametric) treatments for estimating the distribution and transport of compounds between the particle, water and gas phases within the rain-atmosphere system is needed. In particular, the difficulties associated with assigning specific HLC and solubility data to unspecified mixtures from the massive family of PCBs and the only slightly less complex family of dioxins need to be evaluated to provide methods for estimating wet deposition of these important semivolatile compounds.

Additional measurement data are also needed on the less commonly addressed forms of wet deposition such as fog, clouds, and snow. Considerable information has been gathered by NAPAP on the deposition of sulfate and nitrate to vegetation on elevated terrain by fog and clouds. This methodology needs to be extended to TAP deposition, particularly in the region of major stationary TAP sources. Finally, field and laboratory measurements are sorely needed for determining how to best project deposition for snow.

REFERENCES

Banerjee S. and Howard P.H. (1988) Improved estimation of solubility and partitioning through correction of UNIFAC-derived activity coefficients. Environ. Sci. Technol. 22, 839.

Bidleman T.F. (1988) Atmospheric processes: wet and dry deposition of organic compounds are controlled by their vapor-particle partitioning. Environ. Sci. Technol. 22, 361.

Black H.D. and Landsberg H.E. (1983) A method for continuous records of the pH of low-level clouds. Report SR-83-17, University of Maryland, Department of Meteorology.

Bowman C.R. Jr., Geary H.V. Jr. and Schewe G.J. (1987) Incorporation of wet deposition in the Industrial Source Complex Model. Presented at the 80th Annual Meeting of the Air Pollution Control Association, New York, NY.

Brewer D.L., Gordon R.J. and Shepard L.S. (1983) Chemistry of mist and fog from the Los Angeles urban area. Amos. Env., 11, 2267-2270.

Campbell S.A. (1989) Review and comparison of currently recommended methods for computing dry deposition velocity. In: Risk assessment study of the Dickerson site (PPER 1989). Appendix D.

Dollard G.J. and Unsworth M.H. (1983) Field measurements of turbulent fluxes of wind-driven fog drops to a grass surface. Atmos. Env., 17, 775-780.

Engleman, R.J. et al. (1966) Washout coefficients for selected gases and particulates. USAEC Report BNWL-SA-657, Battelle Pacific Northwest Laboratories.

EPA. (1986) Method for the assessment of health risks associated with multiple pathway exposure to municipal waste combustor emissions. Staff paper submitted for review to the Science Advisory Board by the Office of Air Quality Planning and Standards, RTP, North Carolina and the Environmental Criteria and Assessment Office, Cincinnati, Ohio, pp. 3-10 to 3-23.

Falconer R.E. and Falconer P.D. (1979) Determination of cloud water acidity in the Adirondack Mountains of New York State. Atmospheric Science Center Research Publication No. 741, SUNY, Albany, N.Y.

Finlayson-Pitts B.J. and Pitts J.N. (1986) Atmospheric Chemistry John Wiley and Sons, New York.

Gatz D. F. (1977) A review of chemical tracer experiments on precipitation scavenging. Atmos. Env. 11, 945-953.

Hicks B.B., Baldocchi D.D., Meyers T.P., Hosker R.P. Jr. and Matt D.R. (1987) A preliminary multiple resistance routine for deriving dry deposition velocities from measured quantities. Water, Air and Soil Pollution, 36, 311.

Hosker, R.P. (1980) Practical application of air pollution deposition models -- current status, data requirements, and research needs. In: Proceedings of the International Conference on Air Pollutants and their Effects on the Terrestrial Ecosystem, Banff, Alberta Canada, Mary 10-17, 1980. S.V. Krupa and A.H. Legge (eds.), John Wiley and Sons, New York.

Kyriazopoulos B.D., Livadas G.C. and Angouridakis V.E. (1978) Olympus cumulus project. III. Quantitative measurements of water from artificial draining of summer ground-clouds by large cloud-catchers. Meterorologika, V. 67.

Ligocki M.P., Leuenberger C. and Pankow J.F. (1985a) Trace organic compounds in rain - II. Gas scavenging of neutral organic compounds. Atmos. Envir., 19, 1609-1617.

Ligocki M.P., Leuenberger C. and Pankow J.F. (1985b) Trace organic compounds in rain - III. Particle scavenging of neutral organic compounds. Atmos. Envir., 19, 1619-1626.

McMahon T.A. and Denison P.J. (1979) Empirical atmospheric deposition parameters - a survey. Atmos. Env., 13, 571-585.

Moore G.E. and Anderson G.E. (1988) Estimation of aerosol scavenging lifetimes and development of a wet deposition model. Technical Memorandum prepared for Potomac Electric Power Company, SYSAPP-88/199, Systems Applications, Inc.

Pankow J.F.   (1987)   Review and comparative analysis of the theories on partitioning between the gas and aerosol particulate phases in the atmosphere.   Atmos. Env., 21, 2275.

PPER.   (1989)   Risk assessment study of the Dickerson site. Maryland Department of Natural Resources.   Power Plant and Env. Review Div.

Prodi F. and Tampieri F.   (1982)   The removal of particulate matter from the atmosphere: the physical mechanisms. PAGEOPPH 120, 286-325.

Radke L. F., Hobbs P. V. and Eltgroth M. W.,   (1980) Scavenging of aerosol particles by precipitation.   J. App. Met. 14, 715-722.

Schumann T., Zinder B. and Waldvoge A.   (1988)   Aerosol and hydrometeor concentrations and their chemical composition during winter precipitation along a mountain slope: I. Temporal evolution of the aerosol, microphysical and meteorological conditions.   Atmos. Environ., 22, 1443-1459.

Shieh C.M., Wesely M.L. and Hicks B.B.   (1979)   Estimated dry deposition velocities of sulfur over the eastern United States and surrounding regions.   Atmos. Env., 13, 1361.

Slinn W.G.N.   (1984)   "Precipitation scavenging" In: Atmospheric Science and Power Production, Ed by D. Randerson, DOE/TIC-27601, Technical Information Center, Office of Science and Technical Information, U.S. Department of Energy, Chapter 11.

Waldman J.M., Munger J.W, Jacob D.J., Flagan R.C, Morgan J.J. and Hoffmann M.R.   (1982)   Chemical composition of acid fog.   Science, 218, 677-680.

Weston.   (1988)   Health risk assessment for a resource recovery unit in Montgomery County, Maryland.   Report prepared for Montgomery County Department of Environmental Protection by Roy F. Weston, Inc., West Chester, PA. Appendix 6A.

ACCOUNTING FOR DRY DEPOSITION IN INCINERATOR RISK ASSESSMENTS

Curtis C. Travis, Mark W. Yambert

Office of Risk Analysis
Health and Safety Research Division
Oak Ridge National Laboratory
Oak Ridge, TN

INTRODUCTION

The incineration of municipal solid waste (MSW) results in the release of small amounts of hazardous pollutants. Determining the health and environmental risks associated with MSW incineration requires knowledge of the rate at which facility-emitted pollutants will disperse. The deposition of airborne pollutants onto plant, soil, and water surfaces can contribute substantially to human exposure. For example, the deposition of facility-emitted pollutants onto plants and soil surfaces, ingestion of contaminated plants and soil by cattle, and the ingestion of contaminated beef and dairy products can be a significant pathway of human exposure. Travis and Hattemer-Frey (1987) showed that this particular pathway is the major source of human exposure to background levels of dioxin.

Direct measurement of deposition rates ($\mu g/cm^2/sec$) is not practical, and they must therefore be estimated from known air concentrations. The standard approach for this is multiplying ambient ground-level air concentrations times a proportionality constant known as the deposition velocity (cm/sec) (Baldocchi et al., 1987). Deposition velocities used in health risk assessments typically range from 0.1 cm/sec to 3 cm/sec (Shieh et al., 1979; Travis and Hattemer-Frey, 1989).

Historically, deposition velocities have been assumed to be constant and independent of particle size, meteorological conditions, and terrain surrounding the facility (Duan et al., 1988). More recently, the California Air Resources Board (CARB, 1986) developed a particle size-dependent deposition model based on the work of Sehmel and Hodgson (1978). In addition to this model, two other particle size-dependent models exist (Slinn, 1982; Scire et al., 1987). The purpose of this paper is to evaluate the relative effectiveness of these three models along with the classical particle size-independent model developed by Shieh et al. (1986). Comparisons of model-generated deposition velocities and measured values reported in the literature will be made for particles ranging from 0.01 to 30 microns in diameter.

APPROACHES FOR ESTIMATING DRY DEPOSITION

While numerous models exist to estimate pollutant concentrations in air, only a few quantitative models can estimate deposition velocities. These models, which consider the effects of pollutant-specific parameters, such as particle size, properties of the deposition surface, and atmospheric conditions, generally fall into one of two categories. The first category contains models that are based on results of wind tunnel experiments. The primary advantage of these models is that parameters affecting deposition were carefully controlled and monitored during the experiments. The disadvantage of these models is that they are based on the assumption that results obtained in wind tunnel experiments are applicable to outdoor conditions. Conditions in the wind tunnel, such as the maximum size of the turbulent eddies, are likely to differ substantially from field conditions. Of the four deposition models evaluated in this analysis,

Municipal Waste Incineration Risk Assessment
Edited by C.C. Travis, Plenum Press, New York, 1991

53

only the Sehmel and Hodgson (1978) model is based on wind tunnel measurements.

The second group of models are based on observations from field experiments, which were designed to evaluate the deposition of particulates under actual atmospheric conditions to natural surfaces. Thus, unlike models based on wind tunnel data, no error is introduced by applying these models to outdoor situations. The disadvantage of this second group of models, however, is that it is difficult to control and measure all of the parameters that affect the dry deposition of particulates under field conditions.

In general, dry deposition models assume that large particles (i.e., those greater than 10 microns in diameter) are dispersed by gravitational settling at some terminal velocity determined by the Stokes equation. Small particles are assumed to be transferred by both gravitational settling as well as Brownian motion and molecular diffusion.

The Sehmel and Hodgson (1978) model calculates deposition velocity at a given height above the Earth's surface as a function of gravitational settling, wind speed, surface roughness, Monin-Obukhov length, particle Brownian diffusivity, eddy diffusivity, the kinematic viscosity of air, and an empirical coefficient. This coefficient is based on correlations between particle size and density, surface roughness, wind speed, and the Schmidt number established during wind tunnel experiments (Sehmel and Hodgson, 1978).

The Scire et al. (1987) model differs in the assumption that deposition velocity is dependent on the Stokes number. Like the Sehmel and Hodgson (1978) model, the effect of gravitational settling on the deposition of airborne pollutants is considered. This model is unique in that it also uses stability correction functions determined by Shieh et al. (1986).

The Slinn (1982) model also takes into account the effects of the Stokes number and gravitational settling. In addition, the deposition of particulates is assumed to be a function of collection efficiency, which is a combination of the Brownian efficiency of the particles and the impaction and interception efficiencies of the deposition surface.

The model by Shieh et al. (1986) is similar to Sehmel and Hodgson's (1978) formulation except for the fact that deposition is not assumed to be dependent on particle size or density. Also, instead of a coefficient based on wind tunnel correlations, the model relies on stability correction functions for momentum and trace gases calculated from empirical correlations with field data (Wesely and Hicks, 1977). The model does not include algorithms to account for the effects of gravitational settling.

METHODS

Particle size-dependent empirical values of deposition velocities were obtained from the literature (Chamberlain, 1967; Clough, 1975; Davidson and Wu, 1988; Duan et al., 1988; Garland, 1983; Garland and Cox, 1982; Ibrahim et al., 1983; Jost et al., 1986; Katen and Hubbe, 1983; Klepper and Craig, 1975; Little and Wiffen, 1977; Roed, 1987; Sehmel et al., 1984; Sievering, 1983; Wesely and Hicks, 1977; Wesely et al., 1983). Only deposition velocities measured over grass, agricultural land, fields, and similar surfaces were considered. These surfaces were selected because they are characterized by low-growing vegetation that is typically consumed by agricultural animals and humans. Table 1. lists the measured deposition velocities used in this analysis and their corresponding particle sizes.

Ideally, model predictions of deposition velocity will compare well with measured values for a specific set of meteorological conditions and particle size distribution. The empirical data in Table 1, however, lacked detailed descriptions of the atmospheric conditions present when the deposition velocities were measured, making this approach impractical. Instead, model predictions of maximum, minimum, and average deposition velocity for each particle size were obtained for a variety of meteorological conditions. This was achieved by conducting multiple simulations of the model being evaluated using 10 years of hourly meteorological data (five years each from Tampa, Florida, and from Albany, New York). Wind speed values were adjusted according to methods described in US EPA (1980) to reflect wind speed at a height of one meter.

Model input parameters used in this analysis are listed in Table 2. Computerized routines (CARB, 1986) based on the work of McRae (1981) were used to estimate the Monin-Obukhov length (L) from the meteorological data. Brownian diffusivity values were estimated from Sehmel and Hodgson (1978).

RESULTS

Measured deposition velocities for a given particle size exhibit variability spanning about two orders of magnitude. This variability can be attributed to variability in local meteorological conditions at the time the depositions were measured. Minimum measured deposition velocities occur for particles in the 0.1 to 1.0 micron range, with an observed increase in deposition velocity for both larger and smaller particles.

Predicted deposition velocities from the four models versus measured values obtained from the literature are plotted in Figures 1 through 4. The top line in each figure is the maximum deposition velocity predicted by the model for the 10 years of meteorological data considered. The bottom line is the minimum deposition velocity predicted by the model over the considered meteorological conditions, while the middle line represents average predicted deposition velocity.

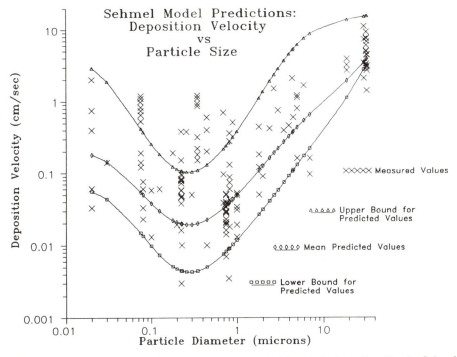

Fig. 1. Upper-bound, lower-bound, and mean deposition velocities (cm/sec) predicted by the Sehmel and Hodgson (1978) model versus measured velocities as a function of particle size.

A comparison of maximum deposition velocities (top line with open triangles) for the three particle size-dependent models (Figures 1 - 3) indicate that they tend to underpredict the upper range of measured velocities for particles 0.05 to one micron in diameter. Of these three models, predictions obtained from the Sehmel and Hodgson (1978) model agree best with measured values. Conversely, Shieh et al.'s (1986) model (Fig. 4) overestimates the upper range of measured deposition velocities for particles in this size range considerably. For particles greater than about two microns in diameter, the opposite becomes true. The Shieh et al. (1986) model estimates of maximum deposition velocities are lower than measured values, while velocities calculated by the Sehmel and Hodgson (1978), Slinn (1982), and Scire et al. (1987) models exceed measured values. This phenomenon is explained by the fact that the latter three models account for the gravitational settling of particles while the Shieh et al. (1986) model does not. For particles larger than one micron, gravitational settling controls deposition. As a result, the Shieh et al. (1986) model tends to underestimate deposition velocity as particle size increases because particle size is not a factor.

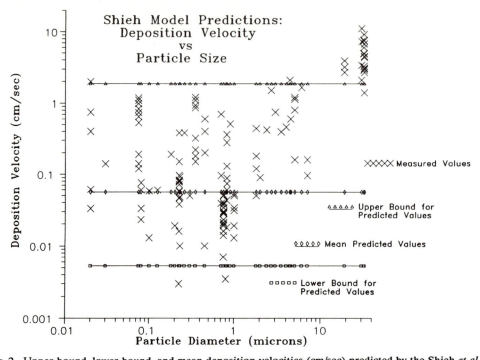

Fig. 2. Upper-bound, lower-bound, and mean deposition velocities (cm/sec) predicted by the Shieh *et al.* (1976) model versus measured velocities as a function of particle size.

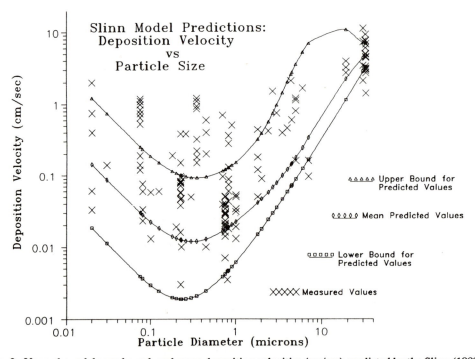

Fig. 3. Upper-bound, lower-bound, and mean deposition velocities (cm/sec) predicted by the Slinn (1982) model versus measured velocities as a function of particle size.

With the exception of predictions made by the Shieh *et al.* (1986) model, predictions of minimum deposition velocity obtained from all four models (bottom line with open squares) are similar and agree relatively well with the lower range of deposition velocities reported in the literature. Minimum deposition velocities calculated by the Shieh *et al.* (1986) model differ by as much as two orders of magnitude from the lowest measured value. Again, for particles greater than a few microns in diameter, this discrepancy is probably due to the lack of a gravitational settling algorithm in the Shieh *et al.* (1986) model. The absence of a gravitational settling term dose not, however, explain why the Shieh *et al.* (1986) model underestimates deposition velocities for particles less than one micron in diameter. This discrepancy is explained by the fact that the other three models evaluated in this analysis assume that resistance to deposition near the surface is primarily governed by diffusion. Since diffusivity varies inversely with particle size, deposition velocities determined by these three models tend to increase with decreasing particle size. In contrast, the Shieh *et al.* (1986) model assumes that near-surface resistance is a function of turbulent and convective mixing (Wesely *et al.*, 1985) and, hence, not strongly influenced by particle size.

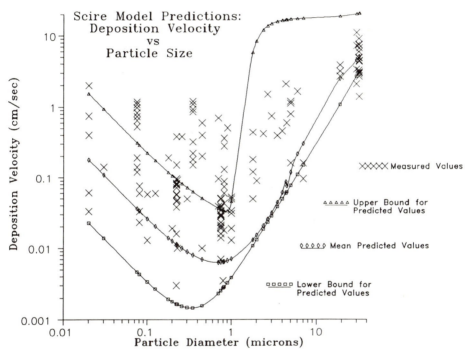

Fig. 4. Upper-bound, lower-bound, and mean deposition velocities (cm/sec) predicted by the Scire *et al.* (1987) model versus measured velocities as a function of particle size.

Of potentially more importance than analysis of maximum and minimum deposition velocities is a comparison of mean measured and predicted values (middle line with open diamonds). Figures 1 through 4 indicate that most measured data lie above the mean values predicted by all four models. A more detailed elucidation of this is shown in Figure 5, which plots mean deposition velocities predicted by each model evaluated against measured values as a function of particle size. This comparison shows that all of the contemporary models tend to underestimate mean deposition velocities for particles in the 0.05 to 1.0 micron diameter range.

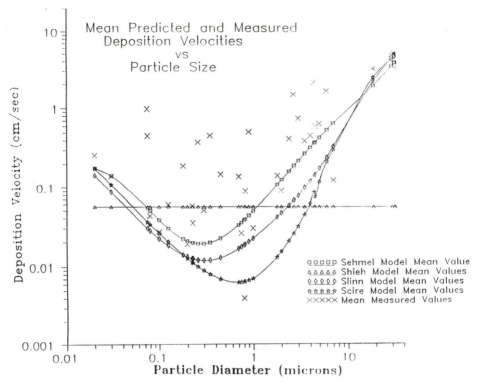

Fig. 5. Mean deposition velocities (cm/sec) predicted all four models evaluated versus measured velocities as a function of particle size.

CONCLUSIONS

Results show that none of the four models evaluated appear to be superior to the others in estimating deposition velocities to agricultural and pasture land. Since deposition models are used primarily for determining the extent of human exposure to facility-emitted pollutants through the food chain, it is desirable to use a model that tends to overestimate rather than underestimate deposition velocities. Using this criterion, we recommend that risk assessors use the Sehmel and Hodgson (1978) model or the model of Shieh *et al.* (1986). With the exception of particles in the 0.05 to 1.0 micron range, deposition velocities predicted by the Sehmel and Hodgson (1978) model are both more conservative and more consistent with measured values. The Shieh *et al.* (1986) model does predict higher values for particle in this size range, but the difference between the Shieh *et al.* 1986 model and measured data is small. Since the Sehmel and Hodgson (1978) model performs as well or slightly better than the other models considered, and it is the current model of choice in scientific community, we believe that it is adequate for estimating particulate deposition velocities to grasslike surfaces.

A model that accurately predicts deposition velocities for all particle sizes and environmental conditions is not yet available. For example, the primary flaw of the Sehmel and Hodgson (1978) model appears to be its inability to accurately predict deposition velocities for particles with a diameter of 0.05 to 1.0 micron. For particles smaller than 0.05 microns and larger than 1.0 micron where deposition velocity is dominated by the well-understood effects of diffusion and gravity, the model performs well. The fact that the models did not accurately predict measured values for particles in this range suggests that the deposition of particles 0.05 to 1.0 microns in diameter is controlled by other process(es) which have not been well characterized. Quantification of this process(es) would improve deposition modeling. Similarly, modifications to account for the effects of diffusion and particle size on deposition velocity would improve the Shieh *et al.* (1986) model.

Table 1
Measured deposition velocities

Particle diameter (microns)	Deposition velocity (cm/sec)	Reference
32.40	3.10	Chamberlain, 1967
32.40	4.40	"
32.40	1.40	"
32.40	4.50	"
32.40	3.00	"
32.40	6.80	"
32.40	2.70	"
32.00	9.20	"
32.00	7.50	"
32.00	6.00	"
32.00	5.00	"
32.00	3.30	"
32.00	2.90	"
32.00	7.60	"
32.00	4.90	"
32.00	3.30	"
19.00	3.90	"
19.00	3.20	"
19.00	2.70	"
5.00	1.20	"
5.00	0.800	"
5.00	0.160	"
2.00	0.090	"
1.00	0.053	"
1.00	0.034	"
1.00	0.019	"
0.08	0.033	"
0.08	0.076	"
0.08	0.023	"
30.00	3.400	Clough, 1975
30.00	7.300	"
30.00	11.100	"
30.00	2.100	"
3.00	0.750	"
3.50	0.395	"
4.00	0.460	"
4.50	0.600	"
0.80	0.0035	Klepper and Craig, 1975
0.70	0.0390	Ibrahim et al., 1983
0.70	0.0960	"
7.000	0.160	"
7.000	0.096	"
4.400	2.100	Cadle et al., 1985[a]
2.700	1.500	"
1.800	0.440	"
0.900	0.510	"
0.300	0.050	Neuman et al., 1985[a]
0.225	0.380	Sievering, 1983
0.225	0.058	Duan et al., 1988
0.225	0.078	"
0.225	0.096	"
0.225	0.057	"
0.225	0.046	"

(continued)

Table 1 (continued)
Measured deposition velocities

Particle diameter (microns)	Deposition velocity (cm/sec)	Reference
0.225	0.072	Duan *et al.*, 1988
0.225	0.016	"
0.225	0.150	"
0.225	0.080	"
0.225	0.003	"
0.225	0.038	"
0.225	0.083	"
0.225	0.082	"
0.225	0.047	"
0.750	0.040	"
0.750	0.029	"
0.750	0.023	"
0.750	0.030	"
0.750	0.034	"
0.750	0.030	"
0.750	0.021	"
0.750	0.050	"
0.750	0.016	"
0.750	0.020	"
0.750	0.014	"
0.750	0.028	"
0.750	0.007	"
0.750	0.051	"
0.750	0.019	"
0.750	0.038	"
0.750	0.040	"
0.075	0.650	Wesely and Hicks, 1977
0.075	1.090	"
0.075	0.740	"
0.075	0.980	"
0.075	1.180	"
0.075	0.830	"
0.075	0.530	"
0.075	0.190	"
0.075	0.140	"
0.075	0.120	"
0.200	0.019	Little and Wiffen, 1977
0.030	0.140	"
0.020	2.000	"
0.125	0.060	Garland and Cox, 1982
1.800	0.050	Garland, 1983
1.800	0.180	"
1.800	0.120	"
1.800	0.120	"
2.400	0.420	"
0.450	0.200	Wesely *et al.*, 1983
0.450	0.600	"
0.450	0.400	"
0.450	0.010	"
0.700	0.700	"
0.820	0.043	Roed, 1987; Jost *et al.*, 1986
0.820	0.018	"
0.820	0.088	"

(continued)

Table 1 (continued)
Table 1 (continued)
Measured deposition velocities

Particle diameter (microns)	Deposition velocity (cm/sec)	Reference
0.820	0.060	Roed, 1987; Jost *et al.*, 1986
0.820	0.074	"
0.820	0.360	"
0.820	0.280	"
0.820	0.130	"
0.350	0.180	"
0.350	0.930	"
0.350	0.860	"
0.350	1.200	"
0.350	1.100	"
0.350	0.320	"
0.350	0.150	"
0.180	0.190	Katen and Hubbe, 1983
0.260	0.380	"
0.020	0.400	Sehmel *et al.*, 1974
0.020	0.750	"
0.020	0.033	"
0.020	0.061	"
0.100	0.013	"
0.100	0.058	"
0.230	0.010	"

[a] Value taken from Davidson and Wu, 1988.

Table 2
Model input data

Variable	Value	Reference
Surface roughness	2.5 cm	Hosker and Lindberg, 1982
Canopy height	0.356 m	"
Zero plane displacement	0.307 m	"
Small collector radius	10.0 microns	Slinn, 1982
Large collector radius	1.0 mm	"
Ratio of viscous to total drag	0.333	"
Fraction of interception by small collectors	0.01	"
Particle density	1.0 g/cm^3	Assumed
Reference height	1.0 m	"

REFERENCES

Baldocchi, D.D., B.B. Hicks, and P. Camara, A canopy stomatal model for gaseous deposition to vegetated surfaces, *Atmos. Environ.*, 21(1): 91-101, 1987.

California Air Resources Board (CARB), *Subroutines for calculating dry deposition velocities using Sehmel's curves*, Prepared by Bart Croes, Sacramento, CA, 1986.

Chamberlain, A.C., Transport of *Lycopodium* spores and other small particles to rough surfaces, *Proc. Royal Acad. Soc.*, 296A, 45-70, 1967.

Clough, W.S., The deposition of particles on moss and grass surfaces, *Atmos. Environ.*, 9, 1113-1119, 1975.

Davidson, C.I., and Y-L. Wu, Dry deposition of particles and vapors, In: *Acid Precipitation, Volume 2. Sources, Emissions, and Modeling*, Adriano, D.C. (Ed), Springer-Verlag, New York, 1988.

Duan, B., C.W. Fairall, and D.W. Thomson, Eddy correlation measurements of the dry deposition of particles in wintertime, *J. Appl. Meteorol.*, 27, 642-652, 1988.

Garland, J.A., Dry deposition of small particles to grass in field conditions, In *Precipitation Scavenging, Dry Deposition, and Resuspension*, H.R. Pruppacher, R.G. Semonin, and W.G.N. Slinn, (Eds), Elsevier, New York, 1983, 849-858.

Garland, J.A., and L.C. Cox, Deposition of small particles to grass, *Atmos Environ.*, 16, 2699-2702, 1982.

Hosker, R.P., Jr., and S.E. Lindberg, Review: Atmospheric deposition and plant assimilation of gases and particles, *Atmos. Environ.*, 16, 889-910, 1982.

Ibrahim, M., L.A. Barrie, and F. Fanaki, An experimental and theoretical investigation of the dry deposition of particles to snow, pine trees and artificial collectors, *Atmos. Environ.*, 17, 781-788, 1983.

Jost, D.T., H.W. Gaggeler, and U. Baltensperger, Chernobyl fallout in size-fractional aerosol, *Nature*, 324, 22-23, 1986.

Katen, P.C., and J.M. Hubbe, Deposition velocity of atmospheric aerosol particles, in *Precipitation Scavenging, Dry Deposition, and Resuspension*, H.R. Pruppacher, R.G. Semonin, and W.G.N. Slinn, (Eds), Elsevier, New York, 1983, 953-962.

Klepper, B., and D.K. Craig, Deposition of airborne particulates onto plant leaves, *J. Environ. Qual.*, 4, 495-499, 1975.

Little, P., and R.D. Wiffen, Emission and deposition of petrol engine exhaust Pb - I. Deposition of exhaust Pb to plant and soil surfaces, *Atmos. Environ.*, 11, 437-447, 1977.

McRae, G.J., Mathematical modeling of photochemical air pollution. Turbulent diffusion coefficients. Ph.D. Thesis, Environmental Engineering Science Department, California Institute of Technology, 1981.

Roed, J., Dry deposition in rural and in urban areas in Denmark, *Radia. Prot. Dosim.*, 21, 33-36, 1987.

Scire, J.S., R.J. Yamartino, D.G. Strimaitis, and S.R. Hanna, *Design for a Non-Steady-State Air Quality Modeling System*, Prepared by Sigma Research Corporation for the California Air Resources Board, Document A025-201, 1987.

Sehmel, G.A., and W.H. Hodgson, A Model for Predicting Dry Deposition of Particles and Gases to Environmental Surfaces, Presented at the 85th American Institute of Chemical Engineers National Meeting, June 4-8, 1978, Philadelphia, PA .

Schmel, G.A., W.H. Hodgson, and S.L. Sutter, Dry deposition of particles, Pacific Northwest Laboratory Annual Report for 1973 to the USAEC Division of Biomedical and Environmental Research, Atmospheric Sciences, BNWL-1850-3, 1974, 157-162.

Shieh, C.M., M.L. Wesely, and C.J. Walcek, *A Dry Deposition Model for Regional Acid Deposition*, EPA/600/3-86/037, Atmospheric Sciences Research Laboratory, Office of Research and Development, U.S. Environmental Protection Agency, Research Triangle Park, North Carolina, 1986.

Shieh, C.M., M.L. Wesely, and B.B. Hicks, Estimated dry deposition velocities of sulfur over the eastern United States and surrounding regions, *Atmos. Environ.*, 13, 1361-1368, 1979.

Sievering, H., Eddy flux and profile measurements of small particle dry deposition velocity at the Boulder Atmospheric Observatory, in *Precipitation Scavenging, Dry Deposition and Resuspension*, Vol. 2, H.R. Pruppacher, R.G. Semonin, and W.G.N. Slinn, Eds., Elsevier, p. 963-977, 1983.

Slinn, W.G.N., Predictions for particle deposition to vegetative canopies, *Atmos. Environ.*, 16, 1785-1794, 1982.

Travis, C.C., and H.A. Hattemer-Frey, Human exposure to dioxin from municipal solid waste incineration, *Waste Manage.*, 9, 151-156, 1989.

Travis, C.C., and H.A. Hattemer-Frey, Human exposure to 2,3,7,8-TCDD, *Chemosphere*, 16: 2331-2342, 1987.

U.S. Environmental Protection Agency (US EPA), *User's Guide for MPTER: A Multiple Point Gaussian Dispersion Algorithm with Optional Terrain Adjustment*, EPA-600/8-80-016, Environmental Sciences Research Laboratory, Research Triangle Park, NC, 1980.

Wesely, M.L., D.R. Cook, R.L. Hart, and R.E. Speer, Measurements and parameterization of particulate sulfur dry deposition over grass, *J. Geophys. Res.*, 90(D1), 2131-2143, 1985.

Wesely, M.L., D.R. Cook, R.L. Hart, B.B. Hicks, J.L. Durham, R.E. Speer, D.H. Stedman, and R.J. Tropp, Eddy-correlation measurements of the dry deposition of particulate sulfur and submicron particles, in *Precipitation Scavenging, Dry Deposition and Resuspension*, Vol. 2, H.R. Pruppacher, R.G. Semonin, and W.G.N. Slinn, Eds., Elsevier, p. 953-962, 1983.

Wesely, M.L., and B.B. Hicks, An eddy correlation measurement of particulate deposition from the atmosphere, *Atmos. Environ.*, 11, 562-563, 1977.

GAS–PARTICLE DISTRIBUTION AND ATMOSPHERIC DEPOSITION

OF SEMIVOLATILE ORGANIC COMPOUNDS

Terry F. Bidleman

Department of Chemistry, Marine Science Program,
and Belle W. Baruch Institute for Marine Biology
and Coastal Research, University of South Carolina,
Columbia, SC 29208, U.S.A.

INTRODUCTION

In the last two decades it has become clear that atmospheric deposition is a major input route of contaminants to the oceans, lakes, and remote continental regions of the world. "Acid rain" is now in school textbooks, but the public is less well acquainted with the aerial transport of metals and toxic organic chemicals. Popular accounts of this problem have appeared in "Discover" magazine (Brown, 1987), as part of a "National Geographic" feature on the Great Lakes (Cobb, 1987), and in the "Toronto Globe and Mail" (Fisher, 1988).

Research into the "atmospheric connection" between toxic substances and the ecosystem falls into three areas: 1. Documentation of evidence that transport and deposition occur. Such evidence consists of direct measurements of metals and organic chemicals in air, precipitation, and dry deposition ("fallout"); contaminant profiles in soil, sediment, and peat bog cores; and the occurrence of anthropogenic organic chemicals in fish and marine mammals from remote lakes and polar regions. 2. Elucidation of the processes responsible for wet and dry deposition, including studies of the chemical and physical forms of contaminants in air and the role of meteorological variables in effecting their removal. 3. Investigation of impacts of atmospherically deposited substances on the ecosystem.

Our work involves the long–range transport and deposition of "semivolatile" organic compounds (SOCs): those with saturation vapor pressures between about 10^{-1} – 10^{-6} pa at ambient temperatures. Included in this group are common pollutants such as pesticides, polychlorinated biphenyls (PCBs), phthalate esters, and polychlorinated dibenzodioxins and furans (PCDDs and PCDFs); as well as compounds having natural and anthropogenic sources: alkanes, polycyclic aromatic hydrocarbons (PAHs), organic acids, and alcohols. Many of these compounds are also sparingly soluble, and as a result they have high octanol–water partition coefficients and accumulate in sediments and in the lipids of aquatic organisms. This article includes a brief review of the atmospheric deposition of SOCs, the distribution of SOCs between the gas and particle phases in ambient air, and the relationship of this distribution to the wet removal of SOCs from the atmosphere.

The literature on atmospheric transport and deposition of organic compounds is extensive, and impractical to cover in this short article. Book chapters, reviews, and proceedings of several recent symposia and workshops dealing with the evidence and processes aspects of atmospheric deposition have been published (Atlas and Giam, 1989; Atlas et al., 1986; Bidleman, 1988; Buat-Menard, 1986; Hites and Eisenreich, 1987; Knap et al., 1990; Kurtz, 1990; Nicholsen, 1988a,b; Strachan and Eisenreich, 1988). These offer a review of the field as well as insights to issues that will be in the forefront in the coming years. Nearly a decade ago, Eisenreich et al. (1981) gave an especially lucid account of toxic substances deposition to the Great Lakes that is an excellent introductory article to the subject.

Contaminant loadings to large lakes and the oceans are, in many cases, dominated by eolean processes. Direct atmospheric input of PCBs to the five Great Lakes lakes was estimated to be 1770 kg/y (Strachan and Eisenreich, 1988). Approximately 60-90% of total PCB loadings to the upper lakes (Superior, Huron, and Michigan) and 6-7% to the lower lakes (Erie, Ontario) were atmospheric. Rain and snow flux of several organochlorine (OC) pesticides to Lake Superior, Cree Lake in northern Manitoba, and Kouchibouguac National Park, New Brunswick was reported by Strachan (1988). Deposition of pollutants with urban sources is especially pronounced near cities. Holsten et al. (1991) found that dry fluxes of PCBs in Chicago were up to three orders of magnitude higher than in nonurban areas. Atlas and Giam (1989) compared atmospheric and riverine inputs of PCBs and OC pesticides to the world's oceans. The contributions of PCBs, chlordane, and DDT compounds by rivers were approximately equal to or 20 times below the lower and upper limits of atmospheric deposition estimates. For dieldrin and the hexachlorocyclohexanes (HCHs), atmospheric loadings were 10 - 260 times greater than those by rivers.

The sedimentary pollution record of remote lakes, bogs, and soils presents a persuasive argument for atmospheric deposition. Profiles of PCBs and OC pesticides in peat bog cores from the Great Lakes region and southeastern Canada closely track their production histories (Rapaport and Eisenreich, 1988). Accumulation rates of PCBs in the sediments of remote Wisconsin lakes were used to estimate atmospheric PCB loadings to Lake Michigan (Swackhamer and Armstrong, 1986,1988; Murphy, 1988). The distribution of PAHs in small forest lakes in Finland was indicative of the development of combustion-related energy production in Finland and nearby countries (Wickstrom and Tolonen, 1987). Soil samples collected from the mid-1800s to the present time from a semirural location in southeast England showed increases in PAH content similar to contemporary atmospheric deposition rates (Jones et al., 1989). Siskewit Lake, a small lake on Isle Royale, Lake Superior that can receive only aerial inputs, has become a field laboratory for atmospheric deposition investigations. Levels of PCDDS and PCDFs in dated sediments from the lake showed that these compounds were virtually absent until after 1940, implying that their source is recent combustion processes (Czuczwa and Hites, 1986). Fish from Siskewit Lake contain a wide variety of OC pesticides and PCBs (Swackhamer and Hites, 1988). A mass balance for PAHs in the lake was constructed from a year's study of PAHs in the air, lake water, and sediments (McVeety, 1986; McVeety and Hites, 1988).

OCs are widespread in fish, birds, and mammals from the Arctic and Antarctic (Addison et al., 1986; Andersson et al., 1988; Kawano et al., 1988; Muir et al., 1988, 1990; Norstrom et al., 1988; Tanabe, 1988), and in lichens and mosses from the Antarctic peninsula (Bacci et al., 1986).

Addison et al. (1986) noted that over the last decade PCBs have declined more rapidly than total DDT in Arctic ringed seal (Phoca hispada) compared to seal populations from eastern Canada. Furthermore, levels of untransformed p,p'-DDT in ringed seal were not significantly different between 1969 and 1981. The authors suggested that DDT input continued for a longer time in the Arctic due to aerial transport from Asia. Muir et al. (1990) found the OC pesticide toxaphene in fish from marine waters and inland lakes across the Northwest Territories. Toxaphene was heavily applied in the southern United States through 1982, and is still used in many other countries of the world. To the best of our knowledge, toxaphene was never used in the Arctic and could only have arrived there by air currents. Atmospheric transport to polar regions and to the open ocean has been confirmed by direct measurements of several OC pesticides and PCBs in air (Atlas and Giam, 1989; Atlas et al., 1986; Bidleman et al., 1989a,b; Hargrave et al., 1988; Kawano et al., 1985; Patton et al., 1989, 1991; Pacyna and Oehme, 1988; Tanabe et al., 1982, 1983) and snow (Bidleman et al., 1989a,b; Gregor and Gummer, 1989; Hargrave et al., 1988; Patton et al., 1989).

Impacts of organic pollutant loadings from the atmosphere have been less well established. Levels of PCBs DDT, and lindane in terrestrial animals in the Scandinavian peninsula were positively correlated to rates of atmospheric deposition (Larsson et al., 1990). In the case of OC pesticides and PCBs, levels in biota from remote areas give rise to concern about long-term effects on animal populations and contamination of our food supply (Tanabe, 1988). Subramanian et al. (1987) observed that testosterone levels in Dall's porpoises (Phocoenoides dalli) from the northwestern Pacific Ocean were negatively correlated with blubber levels of PCBs and DDE. Native people in the Arctic consume large quantities of fish, including their livers. Elevated levels of PCBs have been found in the milk of Inuit women from arctic Quebec (Dewailly et al., 1989).

PHYSICAL STATE OF ORGANIC COMPOUNDS IN THE ATMOSPHERE

SOCs exist in the atmosphere as gases, associated with suspended particles, and dissolved/dispersed in rain, snow, and fog droplets. Sampling techniques must address the need for collecting SOCs in these different phases. The high volume (hi-vol) sampler (Figure 1) in which a glass- or quartz-fiber filter (F) is followed by an adsorbent trap (A) is commonly used to collect particulate and gaseous SOCs in ambient air. Polyurethane foam (PUF) is a suitable adsorbent for SOCs having liquid-phase vapor pressures (p_L^o) less than about 0.1 pa, but adsorbents such as Tenax, XAD-resins, or Florisil must be used for more volatile compounds (Bidleman, 1985; Chuang et al., 1987; Lewis et al., 1982; Zaranski et al., 1991). Hi-vols employing PUF or granular adsorbents have been used to collect PAHs, PCDDs/PCDFs, PCBs, and OC pesticides in ambient air (Chuang et al., 1987; Edgerton et al., 1989; Eitzer and Hites, 1989a; Fairless et al., 1987; and other references in Table 1). Hi-vols are typically operated at flow rates of 0.2 - 0.5 m^3/min, allowing about 300 - 700 m^3/d to be sampled.

The A/F distribution in a hi-vol is often taken as an estimate of the gas-to-particle (G/P) ratio in ambient air. How closely A/F represents G/P is uncertain because of artifacts created by the sampling process. Non-equilibrium effects caused by temperature and concentration changes that occur during the collection period can alter the G/P ratio (Coutant et al., 1988). For example, particle-adsorbed SOCs deposited on the filter at

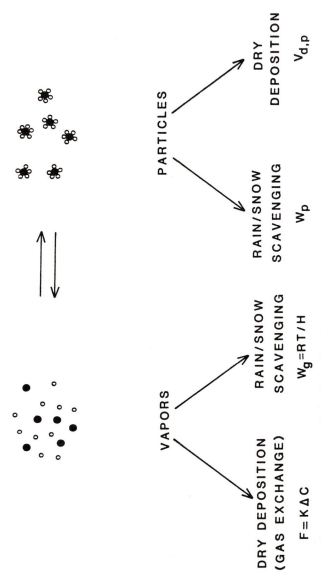

Figure 1. Gas-particle distribution and atmospheric removal processes for SOCs.
Circles: gaseous SOC molecules, black dots: atmospheric particles.

night may be desorbed in the heat of the next day. Another problem, less well recognized than this "blowoff effect", is adsorption of gaseous SOCs by the filter matrix itself. Evidence of this artifact was found by Appel et al. (1989), Cotham (1990), Foreman and Bidleman (1987), Ligocki and Pankow (1989), and McDow and Hutzicker (1990). Diffusion denuders (Figure 1) are being developed with the goal of providing better estimates of the G/P for SOCs (Appel et al., 1989; Coutant et al., 1988; Lane et al., 1988), but at the present time denuder flow rates are only one-tenth or less of hi-vol capabilities.

The G/P distribution of SOCs determines the process of atmospheric deposition: washout of gases and particles, dry deposition of gases and particles, and gas exchange across air-sea and air-lakes surfaces (Figure 2). Reactivities of gaseous organic compounds are quite different than when these compounds are associated with aerosols. Photochemical degradation of PAHs proceeds slowly on fly ash and the PAHs appear to be especially stabilized by adsorption onto ashes of high carbon or iron content (Behymer and Hites, 1988; Dunstan et al., 1989; Yokley et al., 1986).

Some investigations of SOC phase distributions in the atmosphere using the hi-vol are summarized in Table 1. In many of these studies, A/F ratios of individual compounds were correlated to the average sampling temperature (T, kelvin) through a relationship first formulated by Yamasaki et al. (1982):

$$\text{Log } K = \text{Log } A[TSP]/F = m/T + b \qquad (1)$$

where A and F have units of mass SOC/m^3 air, TSP is the total suspended particle concentration ($\mu g/m^3$) and m and b are constants of regression. K is the apparent gas-particle partition coefficient. K has units of mass SOC/m^3 air divided by mass $SOC/\mu g$ particulate matter, or $\mu g/m^3$ -- the same units as TSP. From plots of log K vs 1/T, average values of K at a specific temperature have been derived. These have been correlated to the liquid-phase vapor pressure (p_L^o) of the compound at that temperature:

$$\text{Log } K = m'\text{Log } p_L^o + b' \qquad (2)$$

The basis of equations 1 and 2 and the effect of "non-exchangeable" SOCs on such distributions has been discussed by Pankow (1987, 1988, 1991) and Pankow and Bidleman (1991a,b). The fraction of particulate SOCs (ϕ) can be calculated from A/F by:

$$\phi = 1/[1 + (A/F)] \qquad (3)$$

Hi-vol experiments have thereby provided estimates of ϕ in ambient air, and these have been related to SOC volatilities.

Calculation of ϕ can also be done by a model proposed by Junge (1977) and discussed in detail by Pankow (1987):

$$\phi = c\Theta/(p_L^o + c\Theta) \qquad (4)$$

The parameter c depends on, among other factors, the difference between the heat of desorption from the particle surface and the heat of vaporization of the liquid-phase compound (Pankow, 1987). Junge assumed that c = 17.3

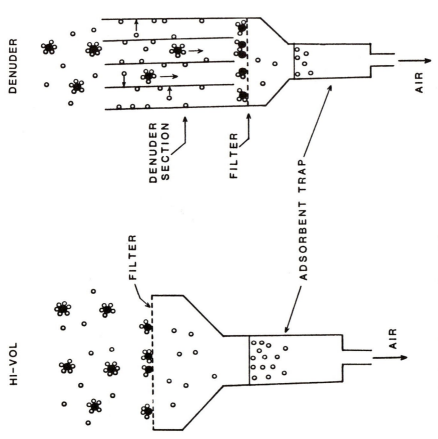

Figure 2. Behavior of gases and particles in a hi-vol and denuder sampler. Gaseous SOCs are removed from the airstream in the denuder section. The sum of SOCs found on the filter and in the back adsorbent trap (desorbed from particles) represents the particulate SOCs in ambient air. Explanations of denuder operation can be found in Appel et al., 1989, Coutant et al. (1988), and Lane et al. (1988).

Table 1. Hi-Vol Investigations of the Phase Distribution of SOCs

<table>
<tr><td colspan="6" align="center">Equation 2 Parameters[a]</td></tr>
<tr><td>Location</td><td>Compounds</td><td>m'</td><td>b'</td><td>r^2</td><td>Reference[b]</td></tr>
<tr><td colspan="6">Urban Air</td></tr>
<tr><td>Osaka, Japan</td><td>PAHs</td><td>1.04</td><td>5.98</td><td>0.99</td><td>1</td></tr>
<tr><td>Columbia, SC</td><td>PAHs</td><td></td><td></td><td></td><td>2</td></tr>
<tr><td>Columbia, SC
Stockholm, Sweden
Denver, CO</td><td>PCBs, OC
pesticides</td><td>0.796</td><td>5.62</td><td>0.96</td><td>3</td></tr>
<tr><td>Portland, OR</td><td>PAHs</td><td>0.882</td><td>5.38</td><td>0.98</td><td>4</td></tr>
<tr><td>Niagara Falls,
Ontario</td><td>PAHs</td><td></td><td></td><td></td><td>5</td></tr>
<tr><td>Columbus, OH</td><td>PAHs</td><td></td><td></td><td></td><td>6</td></tr>
<tr><td>Baltimore, MD</td><td>PAHs</td><td>0.698</td><td>4.94</td><td>0.98</td><td>7</td></tr>
<tr><td>Denver, CO</td><td>PAHs
OC pesticides
alkanes
PCBs
All classes</td><td>0.760
0.925
0.862
0.946
0.830</td><td>5.10
5.63
5.46
5.86
5.35</td><td>0.96
0.99
0.99
0.95
0.97</td><td>8
8
8
8
8</td></tr>
<tr><td>Chicago, IL</td><td>PAHs
PCBs</td><td>0.694
0.726</td><td>4.61
5.18</td><td>0.73
0.69</td><td>9
9</td></tr>
<tr><td>Bloomington, IN</td><td>PCDDs/PCDFs
PCBs</td><td>1.15
0.554</td><td>7.22
4.81</td><td>0.92
0.99</td><td>10
11</td></tr>
<tr><td>Kiel, F.R.G.</td><td>PCBs</td><td></td><td></td><td></td><td>12</td></tr>
<tr><td>Stockholm, Sweden</td><td>PCBs, OC
pesticides</td><td></td><td></td><td></td><td>3,13</td></tr>
<tr><td colspan="6">Rural Air</td></tr>
<tr><td>Coastal Oregon</td><td>PAHs</td><td>0.724</td><td>4.46</td><td>0.98</td><td>4</td></tr>
<tr><td>Lake Superior</td><td>PAHs</td><td>0.586
0.617</td><td>3.83
4.24</td><td>0.95
0.97</td><td>14,15
16</td></tr>
<tr><td>Northern Wis-
consin</td><td>alkanes
PCBs</td><td></td><td></td><td></td><td>17
18</td></tr>
<tr><td>Green Bay, WI</td><td>PAHs</td><td>1.00</td><td>5.47</td><td>0.91</td><td>9</td></tr>
</table>

a) Vapor pressure in pascals. b) References: 1 = Yamasaki et al., 1982; 2 = Keller & Bidleman, 1984; 3 = Bidleman et al., 1986; 4 = Ligocki & Pankow, 1989; 5 = Hoff & Chan, 1987; 6 = Coutant et al., 1988; 7 = Benner et al., 1989; 8 = Foreman & Bidleman, 1990; 9 = Cotham, 1990; 10 = Eitzer & Hites, 1989a; 11 = Hermanson & Hites, 1989; 12 = Duinker & Bouchertall, 1989; 13 = Bidleman et al., 1987; 14 = McVeety, 1986; 15 = McVeety & Hites, 1988; 16 = Baker & Eisenreich, 1990; 17 = Doskey & Andren, 1987; 18 = Manchester-Neesvig & Andren, 1989.

pa–cm and did not vary among compounds. Average values of Θ, the particle surface area per unit volume of air (cm^2/cm^3) are: urban = 1.1×10^{-5}, average background = 1.5×10^{-6}, clean continental background = 4.2×10^{-7} (Bidleman, 1988; Whitby, 1978).

Comparison of Gas–Particle Distribution Estimates

Apparent values of ϕ for PAH and OCs in urban and rural air were derived from hi–vol data presented in several of studies listed in Table 1. Reported A/F ratios were correlated to sampling temperatures through equation 1 and to p_L^o through equation 2, as described by Bidleman et al. (1986) and Foreman and Bidleman (1990). Certain assumptions were made to facilitate comparison of data among the research groups. Some reports did not include TSP values; in these cases TSP concentrations of 60 $\mu g/m^3$ or 30 $\mu g/m^3$ were assumed (10 $\mu g/m^3$ for coastal Oregon, Ligocki and Pankow, 1989), depending on whether the study was done in a city or rural area. A common set of p_L^o values for PAHs as functions of temperature (Yamasaki et al., 1984) was used for equation 2 correlations, regardless of whether these p_L^o values were used in the original article. By this method, slopes and intercepts for equation 2 were obtained from each set of published hi–vol data (Table 1). Slopes (m') were occasionally close to the theoretical value of 1.0, but in other cases m' was quite different from unity. Deviations from unity might be caused by inconstancy in the parameter c (Equation 4) and by kinetic effects when particles re–equilibrate to changing temperatures and vapor concentrations over the sampling period (Pankow and Bidleman, 1991b). Based on the parameters in Table 1, ϕ at two TSP concentrations (100 and 30 $\mu g/m^3$) were calculated.

Estimates of ϕ were also calculated from the Junge–Pankow model (equation 4), using Whitby's average Θ values for urban and average background air, and these results are compared to ϕ from hi–vol sampling in Figure 3. Differences between the model and field results are evident. For example, the predicted ϕ for 2,3,7,8–TCDD in urban air ($p_L^o = 6 \times 10^{-5}$ pa, 25°) is 0.8 from Equation 4, compared to 0.3 from Eitzer's (1989) distributions of other PCDDs and PCDFs, determined with the hi–vol. However the comparisons in Figure 3 lend confidence to the use of equation 4 for predicting ϕ, as was done in two efforts to model wet deposition by the approach described in the following section (Cotham and Bidleman, 1991; Bidleman, 1988; Mackay et al., 1986; Strachan and Eisenreich, 1988).

GAS–PARTICLE DISTRIBUTION AND WET DEPOSITION OF SOCS

Fluxes of SOCs from the atmosphere occur by incorporation of particulate and gaseous SOCs into rain, snow, and fog; dry deposition of particulate SOCs, and exchange of gaseous SOCs across sea and lake surfaces (Bidleman, 1988, Figure 2). The latter process is a two–way street: invasion or evasion of SOCs can occur, depending on whether the water is undersaturated or supersaturated with SOCs relative to the atmosphere. If a disequilibrium exists, fluxes across the air–water interface can be estimated using the two–film model of gas exchange (Mackay and Yuen, 1983). Evidence has been presented to suggest that atmospheric loadings of PCBs to the Great Lakes are episodic, with periods of net deposition followed by outgassing (Eisenreich, 1987; Baker and Eisenreich, 1990).

As an illustration of how the G/P distribution of SOCs governs their removal from the atmosphere, deposition of SOCs by rain is considered here. Concentrations of inorganic and organic substances in rain are often related to those in air through the scavenging ratio, W = Cr/Ca (concentra–

tions on a mass/volume basis). For SOCs, this overall scavenging ratio can be related to scavenging ratios of gases (Wg) and particles (Wp) (Ligocki et al., 1985a,b; Mackay et al., 1986):

$$W = W_g(1-\phi) + W_p\phi \qquad (5)$$

where ϕ and $(1-\phi)$ are the fractions of SOC in the particle and gas phases. If raindrops are in equilibrium with gaseous SOCs, $Wg = RT/H$; H is the Henry's Law constant (pa-m^3/mol) and R is the gas constant (8.3 pa-m^3/mol-deg). Ligocki et al. (1985a) found that low molecular weight PAHs were washed out mainly by vapor dissolution in raindrops, and that field-determined W agreed well with Wg calculated from H values at the rain temperature. Wp is highly variable and is controlled by the particle size distribution and by meteorological factors (Slinn, 1983). A long-term average of $(2-5) \times 10^5$ is suggested by small-particle trace element data (Chan et al., 1986), and an average $Wp = 2 \times 10^5$ was determined for particulate PAHs over Lake Superior (McVeety, 1986).

Table 2. Physical Properties and ϕ for HCHs and p,p'-DDT, 20°C.

	p_L^o, pa	H, pa-m^3/mol	ϕ, Equation 4[a]
α-HCH	6.4×10^{-2}	0.40	3.1×10^{-4}
γ-HCH	3.2×10^{-2}	0.21	8.1×10^{-4}
p,p'-DDT	1.6×10^{-4}	3.5	0.14

a) Average background air, $\Theta = 1.5 \times 10^{-6}$ cm^2/cm^3.

Leuenberger et al. (1988) expressed the particle/dissolved ratio (D) of PAH in rain and snow as:

$$D = W_p c\Theta/S_L^o RT \qquad (6)$$

where c and Θ have been discussed in the context of equation 4 and S_L^o is the liquid-phase water solubility of the PAH. Regression of log D vs. log S_L^o for different rainstorms allowed the quantity WpΘ to be estimated. Assuming $\Theta = 1.1 \times 10^{-5}$ cm^2/cm^3 for urban air (Bidleman, 1988; Whitby, 1978), Wp values calculated from the data of Leuenberger et al. (1988) range from $6.6 \times 10^3 - 2.1 \times 10^5$.

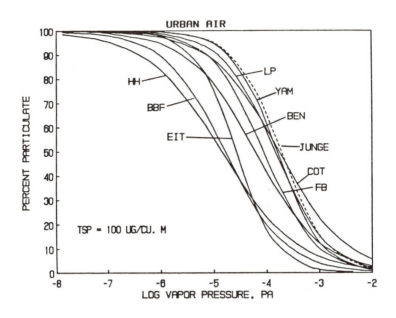

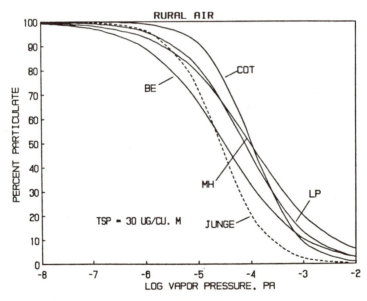

Figure 3. Particulate percentages of SOCs calculated from the Junge-Pankow model (equation 4, dotted line) and estimated by hi-vol sampling (solid lines). Values of Θ (cm^2/cm^3) used in equation 4 were: urban air = 1.1×10^{-5}, average background air = 1.5×10^{-6}. Compounds and references for hi-vol experiments are listed in Table 1. BBF = Bidleman et al. (1986), BE = Baker and Eisenreich (1990), BEN = Benner et al. (1989), COT = Cotham (1990), EIT = Eitzer and Hites (1989a), FB = Foreman and Bidleman (1990), HH = Hermanson and Hites (1989), LP = Ligocki and Pankow (1989), YAM = Yamasaki et al. (1982).

Table 3. Predicted[a] and Field Scavenging Ratios for HCHs and p,p'–DDT

	$W_p\phi$	$W_g(1-\phi)$	W	Dominant[b] Process	Field W[c]
α–HCH	6.2×10^1	6.1×10^3	6.2×10^3	G	$(1.1-3.1) \times 10^4$
γ–HCH	1.6×10^2	1.2×10^4	1.2×10^4	G	$(2.3-3.7) \times 10^4$
p,p'–DDT	2.8×10^4	6.0×10^2	2.9×10^4	P	$(3.3-5.4) \times 10^4$

a) Equation (5), 20°C, $W_p = 2 \times 10^5$.

b) G = gas scavenging, P = particle scavenging

c) Field data from Bidleman and Leonard, 1982; Atlas et al., 1988;
   Ligocki and Pankow, 1985a,b; Bidleman and Christensen, 1979.

The role of physical properties and ϕ in determining the dominent precipitation scavenging mechanism can be seen in the following example for HCHs, and p,p'–DDT. Physical properties of HCHs and p,p'–DDT and equation 4 estimates of ϕ for these pesticides in average background air are given in Table 2. Values of W calculated from equation 5 are compared with field results in Table 3. The agreement is reasonably good, perhaps fortuitously so, considering the widely varied conditions under which the field scaven- ging ratios were determined. The HCHs are washed out mainly by gas dissol- ution in raindrops; particle scavenging plays a minor role. On the other hand, particle removal dominates for DDT, even though only 14% of the DDT is particulate under the conditions of the modeling exercise.

Similar considerations based on equation 5 show the different rainfall removal mechanisms for several SOC types (Figure 4). Gas scavenging con- trols the wet deposition of PAHs having 2-4 rings, the HCHs, and dieldrin. The higher-ring PAHs, PCBs, chlordane, DDT, and n-alkanes are washed out on particles. The relative importance of gas and particle washout for PCDDs and PCDFs was investigated by Eitzer and Hites (1989b). Gas scavenging accounted for 79% of the tetrachloro dibenzofuran removal, but 50% or less for the more highly chlorinated PCDDs and PCDFs. In the absence of gas scavenging, $W = W_p\phi$. This is convincingly shown by a recent study of Duinker and Bouchertall (1989) in which individual PCB congeners were determined in rain and concurrent air samples. The overall scavenging ratios (W) showed a strong correlation with ϕ of the different congeners (estimated by hi-vol sampling), which ranged from <0.01 – 0.3 (Figure 5).

Overall W predicted from equation 5 are compared to field values in Figure 6, and these points can be noted. The data base is very small; only a few simultaneous measurements of SOCs in rain and air have been made. The agreement between actual and predicted W for gas-scavenged compounds is fairly good. Discrepancies may be due to the model assumption of 20°C;

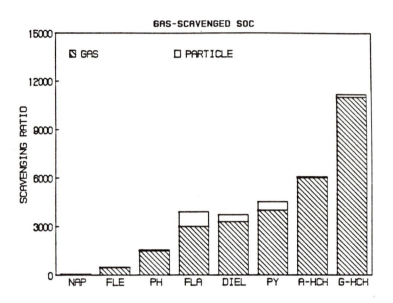

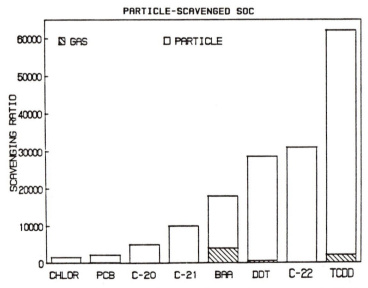

Figure 4. Overall scavenging ratios (W) of SOCs which are removed mainly by gas (top) and particle (bottom) scavenging. In each case the the relative proportions of W contributed by gas (hatched) and particle (open) washout are shown. Calculations were made using equations 4 and 5, for average background air ($\Theta = 1.5 \times 10^{-6}$ cm^2/cm^3), $20^\circ C$, and $Wp = 2 \times 10^5$. Henry's law constants were obtained from several literature sources, and are available from the author upon request. Abbreviations: NAP = naphthalene, PH = phenanthrene, FLA = fluoranthene, PY = pyrene, BAA = benz(a)-anthracene, DIEL = dieldrin, A-HCH and G-HCH = α- and γ-hexachlorocyclohexanes, PCB = polychlorinated biphenyls (Aroclor 1254), CHLOR = chlordane (cis- and trans-chlor-dane), DDT = p,p'-DDT, TCDD = 2,3,7,8-tetrachlorodibenzo-p-dioxin.

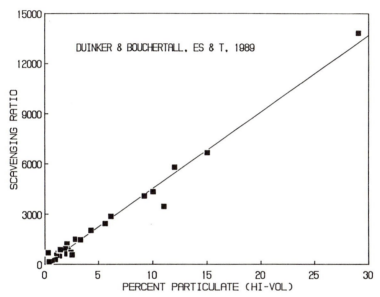

Figure 5. Measured scavenging ratios of PCB congeners vs. the particulate (filter-retained) fraction determined by hi-vol sampling. Data from Duinker and Bouchertall (1989).

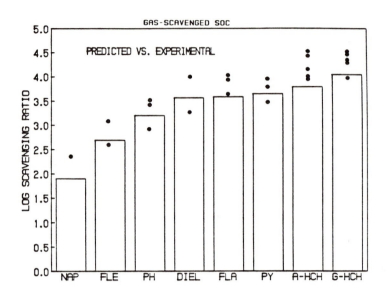

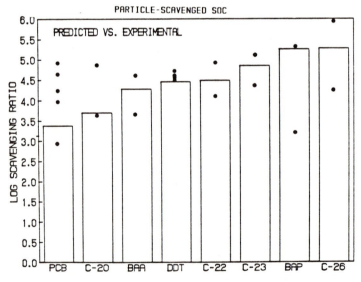

Figure 6. Predicted (equation 5) and experimental scavenging ratios for SOCs removed mainly by gas scavenging (top) and particle scavenging (bottom). Experimental data were taken from Ligocki and Pankow (1985a,b), Farmer and Wade (1986), Atlas and Giam (1981), Atlas et al. (1988), Bidleman and Christensen (1979), Bidleman and Leonard (1982), Murphy (1978), McVeety (1986), Duinker and Bouchertall (1989).

Ligocki et al. (1985a) found that better agreement was obtained by using H values at the actual rain temperature to calculate Wg. The scatter is greater for SOCs removed by particle washout, perhaps a reflection of the large variability in Wp itself from storm-to-storm.

Although the above discussion emphasized an equilibrium approach to gas-particle partitioning and precipitation scavenging, substantial deviations from equilibrium may occur in some systems. SOCs are also concentrated into fog droplets, as shown by recent studies of organophosphate insecticides and herbicides (Schomberg et al., 1991; Glotfelty et al., 1987), hydrocarbons (Leuenberger et al., 1988), and PCDDs/PCDFs (Czuczwa et al., 1989). Concentrations of certain pesticides in filtered fog water were greatly enriched over levels predicted from equilibrium air-water partitioning. Leuenberger et al. (1988) found highly water insoluble n-alkanes in filtered rain, snow, and fog water. These authors and Schomberg et al. (1991) suggested the formation of colloids to explain the apparent enrichment. The elevated concentrations of SOCs in fog droplets coupled with their high deposition velocities might make this vector important for exposing vegetation to toxic chemicals (Glotfelty et al., 1987).

CONCLUSIONS

Wet deposition of PCBs, DDT, PCDDs/PCDFs, and several other SOCs is particle-mediated, and therefore favored by conditions leading to greater ϕ: low temperatures and high concentrations of TSP. Assuming uniform precipitation, one would expect a greater proportion of these SOCs in urban air to be washed out in the winter than the summer. In the Arctic, very low temperatures and the presence of "haze" aerosols in the winter (Barrie, 1986) may tilt the gas-particle equilibrium of SOCs toward the particle side and increase atmospheric loadings to the tundra and the Arctic Ocean (Patton et al., 1991; Cotham and Bidleman, 1991).

Predicting atmospheric fluxes of particle-bound SOCs is limited by uncertainties in gas-particle distributions, plus uncertainties in the particle deposition parameters themselves. Despite artifact problems with the hi-vol, it is encouraging that the filter-retained fraction is similar to ϕ estimated from the Junge-Pankow adsorption model. Continued use of the hi-vol is supported by the fact that the filter-retained fraction of PCB congeners correlates well with their scavenging ratios (Figure 5).

ACKNOWLEDGMENTS

Our work on atmospheric transport, deposition, and physical state of organic compounds in the atmosphere has been supported by the National Science Foundation (Atmospheric Chemistry Division), the U.S. Environmental Protection Agency (Great Lakes Program Office), and the South Carolina Sea Grant Consortium. We have been provided an opportunity to carry out some of these studies in the Canadian Arctic, thanks to the Atmospheric Environment Service, Department of Fisheries and Oceans, and the Polar Continental Shelf Project of Canada. Students participating in this work over several years were: W. Neil Billings, Eric J. Christensen, William E. Cotham, William T. Foreman, Howard W. Harder, Daniel A. Hinckley, Celia D. Keller, John R. Kucklick, Laura L. McConnell, Barnabe Ngabe, Gregory W. Patton, Ross Leonard, and Mark T. Zaranski.

REFERENCES

Addison, R.F., Zinck, M., Smith, T.M. (1986). PCBs have declined more than DDT-group residues in arctic ringed seals (Phoca hispada) between 1972 and 1981. Environ. Sci. Technol. 20, 253–256.

Andersson, Ö, Linder, C.-E., Olsson, M., Reutergårdh, L., Uvemo, U.-B., Wideqvist, U. (1988). Spatial differences and temporal trends of organo-chlorine compounds in biota from the northwestern hemisphere. Arch. Environ. Contam. Toxicol. 17, 755–765.

Appel, B.R., Cheng, W., Salaymeh, F. (1989). Sampling of carbonaceous particles in the atmosphere -- II. Atmos. Environ. 24, 2167–2175.

Atlas, E.L., Bidleman, T.F., Giam, C.S. (1986). Atmospheric transport of PCBs to the oceans. In: Waid, J.S. (ed.) PCBs and the Environment, Vol. I, Chap. 4, CRC Press, Boca Raton, FL, pp. 79–100.

Atlas, E.L., Madero, M., Giam, C.S. (1988). Ambient concentrations and precipitation scavenging of atmospheric organic pollutants. Water, Air, Soil Pollut. 39, 19–36.

Atlas, E.L., Giam, C.S. (1981). Global transport of organic pollutants: ambient concentrations in the remote marine atmosphere. Science 211, 163–165.

Atlas, E.L., Giam, C.S. (1989). Sea-air exchange of high molecular weight synthetic organic compounds: results from the SEAREX program. In: Riley, J.P., Chester, R. (eds.) Chemical Oceanography, Academic Press, NY, 340–378.

Bacci, E., Calamari, D., Gaggi, C., Fanelli, R., Focardi, S., Morosini, M. (1986). Chlorinated hydrocarbons in lichen and moss samples from the Antarctic peninsula. Chemosphere 15, 747–754.

Baker, J.E., Eisenreich, S.J. (1990). Concentrations and fluxes of PAHs and PCB congeners across the air-water interface of Lake Superior. Environ. Sci. Technol. 24, 342–352.

Barrie, L. (1986). Arctic air pollution: an overview of current knowledge. Atmos. Environ. 20, 643–663.

Behymer, T.D., Hites, R.A. (1988). Photolysis of polycyclic aromatic hydrocarbons adsorbed on fly ash. Environ. Sci. Technol. 22, 1311–1319.

Benner, B.A., Gordon, G.E., Wise, S.A. (1989). Mobile sources of atmospheric polycyclic aromatic hydrocarbons: a roadway tunnel study. Environ. Sci. Technol. 23, 1269–1278.

Bidleman, T.F. (1985). High volume collection of organic vapors using solid adsorbents. In: Lawrence, J.F. (ed.) Trace Analysis, Vol. 4, Academic Press, NY, pp. 51–100.

Bidleman, T.F. (1988). Atmospheric processes: wet and dry deposition of organic compounds are controlled by their vapor-particle partitioning. Environ. Sci. Technol. 22, 361–367, correction in 22, 726–727.

Bidleman, T.F., Billings, W.N., Foreman, W.T. (1986). Vapor-particle par-titioning of semivolatile organic compounds: estimates from field collec-tions. Environ. Sci. Technol. 20, 1038–1043.

Bidleman, T.F., Christensen, E.J. (1979). Atmospheric removal processes for high molecular weight organochlorines. J. Geophys. Res. 84, 7857–7862.

Bidleman, T.F., Leonard, R. (1982). Aerial transport of pesticides over the northern Indian Ocean and adjacent seas. Atmos. Environ. 16, 1099–1107.

Bidleman, T.F., Patton, G.W., Hinckley, D.A., Walla, M.D., Cotham, W.E., Hargrave, B.T. (1990). Chlorinated pesticides and polychlorinated biphenyls in the atmosphere of the Canadian Arctic. In: Kurtz, D. (ed.). Long-Range Transport of Pesticides, Lewis Publishers, Chelsea, Michigan, 342–372.

Bidleman, T.F., Patton, G.W., Walla, M.D., Hargrave, B.T., Vass, P., Erickson, P.E., Fowler, B., Scott, V., Gregor, D. (1989). Toxaphene and other organochlorines in Arctic Ocean fauna: evidence for atmospheric delivery. Arctic 42, 307–313.

Bidleman, T.F., Wideqvist, U., Jansson, B., Söderlund, R. (1987). Organochlorine pesticides and PCBs in the atmosphere of southern Sweden. Atmos. Environ. 21, 641–654.

Brown, M.H. (1987). Toxic wind. Discover, Nov. 42–49.

Buat-Menard, P. (1986). The Role of Air–Sea Exchange in Geochemical Cycling, NATO-ASI Series, Vol. 185, D. Reidel, Boston.

Chan, W.H., Tang, A.J.S., Chung, D.H.S., Lusis, M.A. (1986). Concentration and deposition of trace metals in Ontario — 1982. Water, Air, Soil Pollut. 29, 373–389.

Chuang, J.C., Hannen, S.W., Wilson, N.K. (1987). Field comparison of polyurethane foam and XAD-2 resin for air sampling of polynuclear aromatic hydrocarbons. Environ. Sci. Technol. 21, 798–804.

Cobb, C.E., Jr. (1987). The Great Lakes' troubled waters. Nat. Geogr. 172(1), 2–31.

Cotham, W.E., Jr. (1990). Chemical and Physical Processes Affecting the Transport and Fate of Semivolatile Organic Contaminants in the Environment. Ph.D. dissertation, Dept. of Chemistry, University of South Carolina.

Cotham, W.E., Jr., Bidleman, T.F. (1991). Estimating the atmospheric deposition of organochlorine compounds to the Arctic. Chemosphere 22, 165–188.

Coutant, R.W., Brown, L., Cheung, J.C. (1988). Phase distribution and artifact formation in ambient air sampling for polynuclear aromatic hydrocarbons. Atmos. Environ. 22, 403–409.

Czuczwa, J.M., Hites, R.A. (1986). Airborne dioxins and dibenzofurans: sources and fates. Environ. Sci. Technol. 20, 195–200.

Czuczwa, J.M., Katona, V., Pitts, G., Zimmerman, M., DeRoos, F., Capel, P., Giger, W. (1989). Analysis of fog samples for PCDD and PCDF. Chemosphere 18, 847–850.

Dewailly, E., Nantel, A., Weber, J.-P., Meyer, F. (1989). High levels of PCBs in breast milk of Inuit women from arctic Quebec. Bull. Environ. Contam. Toxicol. 43, 641–646.

Doskey, P.V., Andren, A.W. (1986). Particulate and vapor-phase n-alkanes in the northern Wisconsin atmosphere. Atmos. Environ. 20, 1735–1744.

Duinker, J.C., Bouchertall, F. (1989). On the distribution of atmospheric polychlorinated biphenyl congeners between vapor phase, aerosols, and rain. Environ. Sci. Technol. 23, 57–62.

Dunstan, T.D.J., Mauldin, R.F., Jinxlan, Z., Hipps, A.D., Wehry, E.L., Mamantov, G. (1989). Adsorption and photodegradation of pyrene on magnetic, carbonaceous, and mineral subfractions of coal stack ash. Environ. Sci. Technol. 23, 303–308.

Edgerton, S.A., Czuczwa, J.M., Rench, J.D. (1989). Ambient air concentrations of polychlorinated dibenzo-p-dioxins and dibenzofurans in Ohio: Sources and health risk assessment. Chemosphere 18, 1713–1730.

Eisenreich, S.J. (1987). Chemical limnology of PCBs in Lake Superior. In: Hites, R.A., Eisenreich, S.J. (eds.) Sources and Fates of Aquatic Pollutants, Adv. in Chem. Ser., Vol. 216, Amer. Chem. Soc., Washington, DC, pp. 393–469.

Eisenreich, S.J., Looney, B.B., Johnson, T.C. (1981). Airborne contaminants in the Great Lakes ecosystem. Environ. Sci. Technol. 15, 30–38.

Eitzer, B.D., Hites, R.A. (1989a). Polychlorinated dibenzo-p-dioxins and dibenzofurans in the ambient atmosphere of Bloomington, Indiana. Environ. Sci. Technol. 23, 1389–1395.

Eitzer, B.D., Hites, R.A. (1989b). Atmospheric transport and deposition of polychlorinated dibenzo-p-dioxins and dibenzofurans. Environ. Sci. Technol. 23, 1396–1401.

Fairless, B.J., Bates, D.I., Hudson, J., Kleopfer, R.D., Holloway, T.T., Morey, D.A. (1987). Procedures used to measure the amount of 2,3,7,8-TCDD in ambient air near a Superfund site cleanup operation. Environ. Sci. Technol. 21, 550–555.

Farmer, C.T., Wade, T.L. (1986). Relationship of ambient atmospheric hydrocarbon concentrations to deposition. Water, Air, Soil Pollut. 29, 439–452.

Foreman, W.T., Bidleman, T.F. (1990). Apparent vapor-particle distribution of semivolatile organic compounds in Denver, CO. Atmos. Environ. 24A, 2405–2416.

Fisher, M. (1988). Soviet, European pollution threatens health in Arctic. Toronto Globe and Mail, no. 43,378, Dec. 15. pp. A1–A2.

Glotfelty, D.E., Seiber, J.N., Liljedahl, L. (1987). Pesticides in fog. Nature 325, 602–605.

Gregor, D., Gummer, W.D. (1989). Evidence of atmospheric transport and deposition of organochlorine pesticides and PCBs in Canadian Arctic snow. Environ. Sci. Technol. 23, 561–565.

Hargrave, B.T., Vass, W.P., Erickson, P.E., Fowler, B.R. (1988). Atmospheric transport of organochlorines to the Arctic Ocean. Tellus 40B, 480–493.

Hermanson, M.H., Hites, R.A. (1989). Long-term measurements of atmospheric polychlorinated biphenyls in the vicinity of superfund dumps. Environ. Sci. Technol. 23, 1253–1258.

Hites, R.A., Eisenreich, S.J. (1987). Sources and Fates of Aquatic Pollutants, Adv. in Chem. Ser., Vol. 216, Amer. Chem. Soc., Washington, DC, 558 pp.

Hoff, R.M., Chan, K.-W. (1987). Measurement of polycyclic aromatic hydrocarbons in the air along the Niagara River. Environ. Sci. Technol. 21, 556-561.

Holsten, T.M., Noll, K.E., Liu, S.-P., Lee, W.-J. (1991). Dry deposition of polychlorinated biphenyls in urban areas. Environ. Sci. Technol. 25, 1075-1081.

Jones, K.C., Stratford, J.A., Waterhouse, K.S., Furlong, E.T., Giger, W., Hites, R.A., Schaffner, C., Johnson, A.E. (1989). Increases in the polynuclear aromatic hydrocarbon content of an agricultural soil over the last century. Environ. Sci. Technol. 23, 91-105.

Junge, C.E. (1977). Basic considerations about trace constituents in the atmosphere as related to the fate of global pollutants. In: Suffet, I.H. (ed.) Fate of Pollutants in the Air and Water Environments, Part I, Adv. in Environ. Sci. Technol Ser., Vol. 8, Wiley-Interscience, NY, pp. 7-25.

Kawano, M., Tanabe, S., Inoue, T., Tatsukawa, R. (1985). Chlordane compounds found in the marine atmosphere from the southern hemisphere. Trans. Tokyo Univ. Fisheries 6, 59-66.

Kawano, M., Inoue, T., Wada, T., Hidaka, H., Tatsukawa, R. (1988). Bioconcentration and residue patterns of chlordane compounds in marine animals: Invertebrates, fish, mammals, and seabirds. Environ. Sci. Technol. 22, 792-797.

Keller, C.D., Bidleman, T.F. (1984). Collection of airborne polycyclic aromatic hydrocarbons and other organic compounds with a glass fiber filter - polyurethane foam system. Atmos. Environ. 18, 837-845.

Knap, A.H., Atlas, E.L., Church, T.M., Galloway, J.N., Prospero, J.M., Whelpdale, D.M., Kaiser, M.S. (1990). Long-Range Transport of Natural and Contaminant Substances, Kluwer Academic Press, Dordrecht, The Netherlands.

Kurtz, D. (1990). Long-Range Transport of Pesticides, Lewis Publishers, Chelsea, Michigan, 462 pp.

Lane, D.A., Johnson, N.D., Barton, S.C., Thomas, G.H.S., Schroeder, W.H. (1988). Development and evaluation of a novel gas and particle sampler for semivolatile organic compounds in ambient air. Environ. Sci. Technol. 22, 941-947.

Larsson, P., Okla, L., Woin, P. (1990). Atmospheric transport of persistent pollutants governs uptake by holarctic terrestrial biota. Environ. Sci. Technol. 24, 1599-1601.

Leuenberger, C., Czuczwa, J., Heyerdahl, E., Giger, W. (1988). Aliphatic and polycyclic aromatic hydrocarbons in urban rain, snow, and fog. Atmos. Environ. 22, 695-705.

Lewis, R.G., Jackson, M.D. (1982). Modification and evaluation of a high volume air sampler for pesticides and semivolatile industrial organic chemicals. Anal. Chem. 54, 592-594.

Ligocki, M.P., Leuenberger, C., Pankow, J.F. (1985a). Trace organic compounds in rain. II. gas scavenging of neutral organic compounds. Atmos. Environ. 19, 1609–1617.

Ligocki, M.P., Leuenberger, C., Pankow, J.F. (1985b). Trace organic compounds in rain. III. particle scavenging of neutral organic compounds. Atmos. Environ. 19, 1619–1626.

Ligocki, M.P., Pankow, J.F. (1989). Measurement of the gas/particle distributions of atmospheric organic compounds. Environ. Sci. Technol. 23, 75–83.

Mackay, D., Paterson, S., Schroeder, W.H. (1986). Model describing the rates of transfer processes of organic chemicals between atmosphere and water. Environ. Sci. Technol. 20, 810–816.

Mackay, D., Yuen, A.T.K. (1983). Mass transfer coefficients for volatilization of organic solutes from water. Environ. Sci. Technol. 17, 211–217.

Manchester-Neesvig, J.B., Andren, A.W. (1989). Seasonal variations in the atmospheric concentration of polychlorinated biphenyl congeners. Environ. Sci. Technol. 23, 1138–1148.

McDow, S.R., Hutzicker, J.J. (1990). Vapor adsorption artifact in the sampling of organic aerosol: face velocity effects. Atmos. Environ. 24A, 2563–2571.

McVeety, B. (1986). Atmospheric Deposition of Polycyclic Aromatic Hydrocarbons to Water Surfaces: a Mass Balance Approach. Ph.D. dissertation, Dept. of Chemistry, Indiana University, Bloomington, IN.

McVeety, B., Hites, R.A. (1988). Atmospheric deposition of polycyclic aromatic hydrocarbons to water surfaces: a mass balance approach. Atmos. Environ. 22, 511–536.

Muir, D.C.G., Norstrom, R.J., Simon, M. (1988). Organochlorine contaminants in arctic marine food chains: accumulation of specific polychlorinated biphenyls and chlordane-related compounds. Environ. Sci. Technol. 22, 1071–1079.

Muir, D.C.G., Grift, N.P., Ford, C.A., Reiger, Q.W., Hendzel, M.R., Lockhart, W.L. (1990). Evidence for long range transport of toxaphene to remote arctic and subarctic waters from monitoring of fish tissues. In: Kurtz, D. (ed), Long-Range Transport of Pesticides, Lewis Publishers, Chelsea, Michigan, 329–346.

Murphy, T.J. (1978). Polychlorinated biphenyls in precipitation in the Lake Michigan basin. EPA-600/3-78-071, U.S. Environmental Protection Agency, Duluth, MN.

Murphy, T.J. (1988). Comments on "Estimation of the atmospheric and non-atmospheric contributions and losses of polychlorinated biphenyls for Lake Michigan on the basis of sediment records of remote lakes", by D.L. Swackhamer and D.E. Armstrong. Environ. Sci. Technol. 22, 230.

Nicholsen, K.W. (1988). A review of particle resuspension. Atmos. Environ. 22, 2639–2651.

Nicholsen, K.W. (1988). The dry deposition of small particles: a review of experimental measurements. Atmos. Environ. 22, 2653–2666.

Norstrom, R.J., Simon, M., Muir, D.C.G., Schweinsburg, R.E. (1988). Organochlorine contaminants in arctic marine food chains: identification, geographical distribution, and temporal trends in polar bears. Environ. Sci. Technol. 22, 1063–1071.

Pacyna, J., Oehme, M. (1988). Long-range transport of some organic compounds to the Norwegian Arctic. Atmos. Environ. 22, 243–257.

Pankow, J.F. (1987). Review and comparative analysis of the theories on partitioning between the gas and aerosol particulate phases in the atmosphere. Atmos. Environ. 21, 2275–2283.

Pankow, J.F. (1988). The calculated effects of non-exchangeable material on the gas-particle distributions of organic compounds. Atmos. Environ. 22, 1405–1409.

Pankow, J.F. (1991). Common Y-intercept and single compound regressions of gas-particle partition coefficient vs. 1/T. Atmos. Environ. (in press).

Pankow, J.F., Bidleman, T.F. (1991a). Effects of temperature, TSP, and percent non-exchangeable meterial in determining the gas-particle distributions of organic compounds. Atmos. Environ. (in press).

Pankow, J.F., Bidleman, T.F. (1991b). Interdependence of the slopes and intercepts from log-log correlations of measured gas-particle partitioning vs. subcooled liquid vapor pressure. 1. Basic considerations and review of available data. Atmos. Environ. (in press).

Patton, G.W., Hinckley, D.A., Walla, M.D., Bidleman, T.F., Hargrave, B.T. (1989). Airborne organochlorines in the Canadian High Arctic. Tellus 41B, 243–255.

Patton, G.W., Bidleman, T.F., Barrie, L.A. (1991). Polycyclic aromatic and organochlorine compounds in the atmosphere at northern Ellesmere Island, Canada. J. Geophys. Res. 96, 10,867–10,877.

Rapaport, R.A., Eisenreich, S.J. (1988). Historical atmospheric inputs of high molecular weight chlorinated hydrocarbons to eastern North America. Environ. Sci. Technol. 22, 931–941.

Schomberg, S., Glotfelty, D.E., Seiber, J.N. (1991). Pesticide occurrence and distribution in fog collected near Monteray, California. Environ. Sci. Technol. 25, 155–160.

Slinn, W.G.N. (1983). Air-to-sea transfer of particles. In: Liss, P.S., Slinn, W.G.N. (eds.) Air-Sea Exchange of Gases and Particles, D. Reidel, Dordrecht, Holland, pp. 299–405.

Strachan, W.M.J. (1988). Toxic contaminants in rainfall in Canada: 1984. Environ. Toxicol. Chem. 7, 871–877.

Strachan, W.M.J., Eisenreich, S.J. (1988). Mass balancing of toxic chemicals in the Great Lakes: the role of atmospheric deposition. Report to the International Joint Commission of the workshop on estimation of atmospheric loadings of toxic chemicals to the Great Lakes, Scarborough, Ont., Oct. 1986.

Subramanian, A., Tanabe, S., Tatsukawa, R., Saito, S., Miyazaki, N. (1987). Reduction in the testosterone levels by PCBs and DDE in Dall's porpoises of northwestern North Pacific. Mar. Pollut. Bull. 18, 643–646.

Swackhamer, D.L., Armstrong, D.E. (1986). Estimation of the atmospheric and nonatmospheric contributions and losses of polychlorinated biphenyls for Lake Michigan on the basis of sediment records of remote lakes. Environ. Sci. Technol. 20, 879–883.

Swackhamer, D.L., Armstrong, D.E. (1988). Reply to comments of T.J. Murphy. Environ. Sci. Technol. 22, 230–231.

Swackhamer, D.L., Hites, R.A. (1988). Occurrence and bioaccumulation of organochlorine compounds in fishes from Siskewit Lake, Isle Royale, Lake Superior. Environ. Sci. Technol. 22, 543–548.

Tanabe, S. (1988). PCB problems in the future: foresight from current knowledge. Environ. Pollut. 50, 5–28.

Tanabe, S., Tatsukawa, R., Kawano, M., Hidaka, H. (1982). Global distribution and atmospheric transport of chlorinated hydrocarbons: HCH (BHC) isomers and DDT compounds in the western Pacific, eastern Indian, and Antarctic Oceans. J. Oceanog. Soc. Japan 38, 137–148.

Tanabe, S., Hidaka, H., Tatsukawa, R. (1983). PCBs and chlorinated hydrocarbon pesticides in the antarctic atmosphere and hydrosphere. Chemosphere 12, 277–288.

Whitby, K.T. (1978). The physical characteristics of sulfate aerosols. Atmos. Environ. 12, 135–159.

Wickstrom, K., Tolonen, K. (1987). The history of airborne polycyclic aromatic hydrocarbons (PAH) and perylene as recorded in dated lake sediments. Water, Air, Soil Pollut. 32, 155–175.

Yamasaki, H., Kuwata, K., Miyamoto, H. (1982). Effects of ambient temperature on aspects of atmospheric polycyclic aromatic hydrocarbons. Environ. Sci. Technol. 16, 189–194.

Yamasaki, H., Kuwata, K., Kuge, Y. (1984). Determination of vapor pressures of polycyclic aromatic hydrocarbons in the supercooled liquid phase and their adsorption on airborne particulate matter. Nippon Kagaku Kaishi 8, 1324–1329 (in Japanese, Chem. Abst. 101, 156747p)

Yokley, R.A., Garrison, A.A., Wehry, E.L., Mamantov, G. (1986). Photochemical transformation of pyrene and benzo(a)pyrene vapor-deposited on eight coal stack ashes. Environ. Sci. Technol. 20, 86–90.

Zaranski, M.T., Patton, G.W., McConnell, L.L., Bidleman, T.F., Mulik, J.D. (1991). Collection of nonpolar organic compounds from ambient air using polyurethane foam - granular adsorbent sandwich cartridges. Anal. Chem. 63, 1228–1232.

AN OVERVIEW OF FOOD CHAIN IMPACTS FROM MUNICIPAL

WASTE COMBUSTION

Holly A. Hattemer-Frey

Lee Wan & Associates, Inc./Radian
120 South Jefferson Circle
Suite 100
Oak Ridge, TN 37830

Curtis C. Travis

Office of Risk Analysis
Health and Safety Research Division
Oak Ridge National Laboratory
Oak Ridge, TN 37831-6109

INTRODUCTION

Human exposure to pollutants emitted from a municipal waste combustor (MWC) can occur via inhalation, ingestion of contaminated food items, infant consumption of mother's milk, and dermal absorption. Of particular concern, however, are potential exposures from ingesting contaminated food items. The food chain has been shown to be the primary source of human exposure to a large class of organics, including DDT, dioxin, pentachlorophenol, benzo(a)pyrene, and most pesticides (Beck *et al.*, 1989; Hattemer-Frey and Travis, 1989b; Lioy *et al.*, 1988; Travis and Arms, 1987; Travis and Hattemer-Frey, 1987).

Various measurement and predictive techniques can be used to evaluate the movement and transfer of chemicals within and between environmental media as well as the concentration of organics to which humans are exposed. Since organic chemicals tend to accumulate in the media in which they are most soluble, a few basic physicochemical properties can be used to predict the behavior and fate of chemicals released to the environment. Multimedia transport models estimate the concentration of a pollutant in various environmental media and then use those concentrations to predict the amount of pollutant to which humans are exposed.

This paper quantifies the extent of human exposure to 2,3,7,8-tetrachlorinated dibenzo-p-dioxin (TCDD, commonly referred to as dioxin) and cadmium emitted from a typical MWC in the U.S. It also provides an innovative perspective on human exposure to facility-emitted pollutants using a probabilistic risk assessment (PRA) approach. Probabilistic simulation techniques allow more realistic exposure and risk estimates, reduce reliance on worst-case assumptions, and offer a scientifically defensible method of setting emission standards for MWCs, because they take into account variability in facility design, location, and population distributions around existing MWCs and use actual data versus hypothetical extremes.

UPTAKE OF POLLUTANTS IN THE TERRESTRIAL FOOD CHAIN

Assessing the magnitude of exposure to pollutants emitted from an MWC depends largely on predicting their bioaccumulation in the food chain. The octanol-water partition coefficient (K_{ow}) is used extensively to estimate the bioconcentration potential of pollutants in biological systems. Chemicals with large K_{ow} values, or with lipophilic compounds, such as TCDD, DDT, and polychlorinated biphenyls (PCBs), are most soluble in organic matter. These chemicals tend to sorb strongly to air particulates, soil, and sediment; bioaccumulate in living organisms; and transfer to humans through the food chain. Conversely, chemicals with small K_{ow} values, such as trichloroethylene, tetrachloroethylene, and benzene, tend to partition mostly into air or water. In this case, inhalation is the primary pathway of human exposure.

THE TERRESTRIAL FOOD CHAIN MODEL

The Terrestrial Food Chain (TFC) model, developed by Oak Ridge National Laboratory's Office of Risk Analysis, is a multimedia transport model designed to assess the magnitude of human exposure to compounds emitted from a point source (in this case, an MWC). The TFC model offers improvements over existing food chain models, such as those generated by the U.S. Environmental Protection Agency (EPA) (1986b) and Travis *et al.* (1986). Methodologies described in these earlier models were modified to account for the effects of (1) direct deposition of pollutants onto plant, water, and soil surfaces and (2) air-to-leaf transfer of organic vapors. The TFC model uses annual atmospheric deposition rates ($\mu g/m^2$/year) and average annual atmospheric air concentration values ($\mu g/m^3$) to estimate the amount of pollutant entering the terrestrial food chain and the average daily intake of pollutant by individuals living near an MWC.

Deposition rates are important because they define the rate airborne pollutants precipitate from the atmosphere onto soil, water, and plant surfaces. Deposition of airborne pollutants can indirectly contribute to human exposure. For example, the deposition of contaminants onto pastures and grasslands, the accumulation of contaminants in beef and cow milk fats, and the ingestion of contaminated beef and dairy products can be a significant pathway of human exposure to environmentally released pollutants. Travis and Hattemer-Frey (1987) showed that this pathway is the major source of human exposure to background levels of TCDD. The TFC model quantifies the amount of pollutant entering the terrestrial food chain and the extent of human exposure from inhalation and from ingestion of contaminated soil and agricultural products (fruits, grains, vegetables, meats, milk, and eggs).

Accumulation in Vegetation

The accumulation of facility-emitted compounds in vegetation is a complex process that involves root uptake, foliar uptake (i.e., air-to-leaf transfer), and deposition.

<u>Root Uptake.</u> Root uptake of organics has been correlated with K_{ow} (Baes, 1982; Briggs *et al.*, 1982) and can be estimated from B_v, the soil-to-plant bioconcentration factor (BCF). B_v is defined as the equilibrium concentration of pollutant in plant tissue [$\mu g/g$ dry weight (DW)] divided by the equilibrium concentration of pollutant in soil ($\mu g/g$ DW) and can be estimated from the following geometric mean regression equation developed by Travis and Arms (1988) in the absence of measured data:

$$B_v = 38.73 \times K_{ow}^{-0.578} \qquad (1)$$

[Halfon (1985) recommended that a geometric mean regression be used to determine the relationship between K_{ow} values and various measures of bioaccumulation, since both of these parameters are usually measured with error. The traditional linear regression method assumes that one variable is measured with error and the second is known precisely.]

The organic matter content of soil is the single most important factor affecting root uptake of organic pollutants (Nash, 1974). Because lipophilic compounds are most soluble in organic matter, they are likely to sorb strongly to soil, and root uptake is not expected to be a major pathway of vegetative contamination. Briggs *et al.* (1982) showed that root uptake decreases as K_{ow} increases. Conversely, nonlipophilic compounds are not very soluble in organic matter. For these compounds, root uptake is expected to be a dominant pathway of vegetative contamination.

The concentration of pollutant in vegetation (DW basis) due to root uptake (CVR) can be estimated from the following equation:

$$\text{CVR } (\mu g/g \text{ DW}) = B_v \times C_s \qquad (2)$$

where B_v equals the plant-specific, chemical-specific, soil-to-root BCF ($\mu g/g$ plant DW $\div \mu g/g$ soil DW), and C_s is the equilibrium concentration of pollutant in soil ($\mu g/g$ DW). C_s is estimated from:

$$C_s(\mu g/g) = \frac{D_y \times [1.0 - \exp\,(-K_s \times T_c)]}{Z \times D_s \times K_s \times K_1} \qquad (3)$$

where:
D_y = annual atmospheric deposition rate ($\mu g/m^2/year$)
K_s = pollutant-specific soil loss constant (year^{-1})
T_c = length of long-term pollutant build-up in the soil (65 years)
Z = soil mixing depth (20 cm)
D_s = density of soil (1.4 g/cm^3)
K_1 = 100^2 to convert cm^2 to m^2

The TFC model assumes that deposition of facility-emitted pollutants occurs continuously throughout the operating lifetime of the facility. The parameter T_c reflects this long-term pollutant build-up in the soil due to deposition. EPA (1986b) estimated that the lifetime of an MWC ranges from 30 to 100 years. An average value of 65 years was used in this analysis.

The soil loss constant (K_s) reflects the soil's loss of organics due to leaching, degradation, and volatilization. Since metals do not decay or degrade in the environment, K_s values for metals equal 0. The TFC model also assumes that crops are grown in tilled soil and that pollutants are distributed evenly within this soil to a depth of 20 cm. To model direct soil ingestion, soil is considered to be untilled. In this case, effective depth to which pollutants are distributed is assumed to be 1 cm.

Air-to-Leaf Transfer. While some organics are taken up by plants directly from contaminated soil and translocated to upper plant parts, other compounds volatilize from the soil, and their vapors are absorbed by aerial plant parts (Beall and Nash, 1971; Bacci and Gaggi, 1985, 1986, 1987; Bacci et al., 1989). Volatilization of organics from polluted soil may be as or more important than root uptake as a source of plant contamination. Buckley (1982), Nash and Beall (1980), and Mosbaek et al. (1988) showed that PCB, DDT, and mercury contamination of plant foliage was mainly from volatilization of these compounds from contaminated soil.

The accumulation of volatilized organics by upper plant parts can be estimated from B_{va}, the air-to-plant BCF, which is defined as the equilibrium concentration of organic in upper plant parts (μg/g DW) divided by the equilibrium concentration of organic in air *as a vapor* (μg/g). B_{va} is best defined as K_{lw}/K_{aw}, where K_{lw} and K_{aw} represent the leaf-to-water and air-to-water partition coefficients, respectively. Intuitively, K_{lw} should be related to K_{ow}, and K_{aw} to Henry's Law Constant (H) in atm-m^3/mol (Travis and Hattemer-Frey, 1988). Thus, B_{va} should be positively correlated with K_{ow} divided by H. Given this relationship, lipophilic compounds must have a low H value to be accumulated via this pathway.

B_{va} can be estimated from the following geometric mean regression equation based on the experimental data collected by Bacci et al. (1989) for 10 chemicals with varying physicochemical properties:

$$\text{Log } (H \times B_{va}) = -1.15 + 1.19 \text{ Log} \times K_{ow} \tag{4}$$

Mackay (1982) and Halfon (1985) argued that BCFs should be linearly related to K_{ow}, and Travis and Arms (1988) found that the relationship between biotransfer factors (BTFs) for cow milk and beef were also linearly related to K_{ow}. Equation 4 indicates that this same relationship holds true for the bioconcentration of organics by aerial plant parts. We thus constrained the slope in equation 4 to be unity and refitted the data to yield:

$$B_{va} = 5.0 \times 10^{-6} \times K_{ow} \div H \tag{5}$$

Given the limited data, it is impossible to determine whether organics are adhering to the leaf cuticle (i.e., accumulation on the surface only) or are actually being incorporated into plant tissue. Buckley (1982) reported that the uptake of organics by aerial plant parts was a species-specific phenomenon, indicating that some types of vegetation may accumulate higher concentrations of organics than other species.

To calculate the organic concentration in vegetation from air-to-leaf transfer (CVA), the model assumes that organics deposited onto the soil volatilize. The concentration of pollutant in air from volatilization (C_{av} in $\mu g/m^3$) is determined by assuming that a uniform amount of pollutant is emitted from the soil per unit area. A Gaussian plume model was used to estimate the loss of pollutant as it is transported away from the air-soil-water interface to plant height. This model assumes that the area over which the pollutant disperses is essentially circular and takes into account average wind speed, plant height, and a Gaussian plume dispersion coefficient. Thus, the concentration in air due to volatilization of pollutants from contaminated soil can be estimated from the following equation:

$$C_{av} = \sum_{i=1}^{6} F_i \int_{0}^{R} \frac{2q}{\sqrt{2}\pi\,\sigma_{zi}\,u} \cdot \exp\left[-\tfrac{1}{2} \{\frac{Ht}{\sigma_{zi}}\}^2 \right] dx \qquad (6)$$

where: C_{av} = partial concentration of pollutant in air due to volatilization from soil (μg pollutant/m^3 air)

F_i = fraction of time of the ith atmospheric stability

R = radius of the contaminated area (m)

q = volatilization rate of pollutant from soil ($\mu g/m^2/s$)

H = height of plant above soil (m)

u = average wind speed above contaminated area (m/s)

σ_{zi} = Gaussian plume dispersion parameter for the vertical direction (m) for the ith Pasquill atmospheric stability category

The Gaussian plume dispersion parameter for the vertical direction (m) for the ith Pasquill atmospheric stability category, σ_{zi}, is calculated according to the formula

$$\sigma_{zi}(m) = ai\,(x\,/\,K_3)^{bi} \qquad (7)$$

where ai and bi are empirically determined parameters for each Pasquill atmospheric stability category (EPA, 1986a), x is the downwind distance from the source (m), and K_3 is a units conversion factor of 1000.

Thus, the concentration of pollutant on vegetation due to air-to-leaf transfer (CVA) can be estimated from the following equation:

$$CVA\ (\mu g/g\ DW) = \frac{[C_{av} + (C_a \times F_v)] \times B_{va}}{D_a} \qquad (8)$$

where: C_{av} = pollutant concentration in air due to revolatilization ($\mu g/m^3$ air)

C_a = pollutant concentration in air ($\mu g/m^3$)

F_v = fraction of pollutant in the vapor phase

B_{va} = plant-specific, chemical-specific, air-to-leaf BCF (μg/g plant DW ÷ μg/g in air as a vapor)

D_a = density of air (1190 g/m^3)

Junge (1977) theorized that compounds with a vapor pressure greater than 10^{-6} mm Hg will exist predominately in the vapor phase, while chemicals that have a vapor pressure lower than 10^{-6} mm Hg will most likely sorb onto particulate matter. If data on the percentage of compound that exists in the vapor phase are not available in the literature, the following simple sorption equation proposed by Cupitt (1980) can be used to estimate a chemical's partitioning between the gaseous and particulate phases:

$$F_p = (K/P) \div (1 + K/P) \tag{9}$$

where F_p equals the fraction sorbed to particulates, K is a group of constants that characteristically range from 10^{-6} to 10^{-7} mm Hg, and P is the chemical's vapor pressure in mm Hg (Mackay *et al.*, 1986). As a consequence, chemicals with a vapor pressure between 10^{-6} and 10^{-7} will partition equally between the gaseous and the sorbed (particulate) phase (Junge, 1977; Seiber *et al.*, 1982).

It was assumed that protected produce, including potatoes and other root vegetables, legumes, and garden fruits, whose edible portions either grow underground or are protected by pods, shells, or nonedible peels, were not contaminated via this pathway. If accumulation via foliar uptake is a surface phenomenon only (i.e., organics are not being incorporated into plant tissue), this assumption is valid. If, on the other hand, organic vapors are being incorporated into plant tissue through the leaves and other aerial plant parts, then this assumption may not be valid. Metals do not volatilize from the soil and do not exist in air as a vapor. Therefore, with the exception of mercury (de Tammerman *et al.*, 1986; Mosbaek *et al.*, 1988), air-to-leaf contamination for metals is equal to 0.

Deposition. Deposition of pollutants onto outer plant surfaces can also contribute substantially to vegetative contamination, especially for lipophilic compounds, which are expected to sorb strongly to particulates. The concentration of pollutant on vegetation due to atmospheric deposition (CVD, μg/g DW) can be estimated using the following equation:

$$CVD \ (\mu g/g \ DW) = \frac{D_y \times r \times [1.0 - \exp(-K_e \times T)]}{Y \times K_e} \tag{10}$$

where: D_y = annual pollutant deposition rate (μg/m^2/year)

r = intercept fraction for the edible plant portion (unitless);

K_e = weathering half-life (year^{-1})

T = length of the growing season (year)

Y = yield or standing crop biomass of the edible plant portion (g DW/m^2)

The TFC model assumes that deposition occurs only during the growing season and that protected produce is not contaminated via this pathway. The length of the growing season for food crops consumed by humans (T_h) and for grains (T_g) was set at 58 days (0.16 year), while T for forage (T_f) equals 51 days (0.12 year) (Baes et al., 1984). Thus, the value of 58 days represents the longest time required for any food crop to mature.

K_e can be estimated from the following equation:

$$K_e \text{ (year}^{-1}) = \ln 2 / T_{1/2} \tag{11}$$

where ln 2 is the natural log of two and $T_{1/2}$ is the half-life of the compound on the plant surface (year). In the absence of measured data, it is generally assumed that organics deposited onto outer plant surfaces have a 14-day weathering half-life (Baes et al., 1984).

Intercept fraction (r) and yield (Y) values account for the fact that not all airborne pollutants deposited within a given area actually accumulate on edible plant parts. Vegetation-specific differences in the intercept fraction-to-productivity ratio for plants consumed by animals and humans (Tables 1 and 2) require that different estimates of CVD be made for forage, grain, and food crops consumed by humans (Baes et al., 1984). The total concentration of pollutant on vegetation consumed by animals and humans is determined by summing the contribution from each of the three pathways of vegetative contamination (CVR + CVD + CVA).

Accumulation in Animal Products

Assessing the environmental fate of chemicals depends largely on the ability to predict the extent to which they will bioaccumulate in living organisms, including animals and humans. The traditional measure of a chemical's potential to accumulate in biota is the BCF, which is defined as the equilibrium concentration of pollutant in an organism or tissue (μg/kg) divided by the equilibrium concentration of pollutant (μg/kg) in food. This concept, however, is not readily applied to humans and other terrestrial organisms, because the amount of chemical in various food items can vary markedly. Thus, for risk assessment purposes, it is more effective to examine the BTF, which is defined as the equilibrium concentration of pollutant in an organism or tissue (μg/kg) divided by the average daily intake of pollutant (μg/day) (Travis and Arms, 1988).

In the absence of measured data, BTFs for pollutants in cow milk (B_m) and beef (B_b) (whole-tissue basis) can be estimated from the following geometric mean regression equations developed by Travis and Arms (1988):

$$B_b \text{ (day/kg)} = 2.51 \times 10^{-8} \times K_{ow} \quad (n = 36; \ r = 0.81) \tag{12}$$

$$B_m \text{ (day/kg)} = 7.94 \times 10^{-9} \times K_{ow} \quad (n = 28; \ r = 0.74) \tag{13}$$

Table 1. Mean intercept fractions (r) for
vegetation consumed by animals and humans

Plant group	Mean	Range
Forage (r_f)	0.47	0.02–0.82
Grains (r_g)	0.00	–
Potatoes (r_p)	0.00	–
Leafy vegetables (r_{lv})	0.16	0.08–0.38
Legumes (r_{le})	0.008	0.005–0.01
Root vegetables (r_r)	0.00	–
Garden fruits (r_{gf})	0.05	0.004–0.08

Sources: Baes *et al.*, 1984; Hoffman and Baes, 1979.

Measured concentrations of organics in cow milk or beef fat can be converted to a fresh-meat or whole-milk basis assuming that beef and milk are 25% and 3.7% fat, respectively. Daily intake was computed assuming an average feed ingestion rate of 16 kg per day for lactating cows and 8 kg per day for nonlactating cows (Travis and Arms, 1988).

Table 2. Yield (Y) or standing crop biomass of vegetation
consumed by humans and animals

Vegetation type	Mean (g DW/m^2)	Range (g DW/m^2)
Forage	310	20–750
Grains	300	140–450
Potatoes	480	410–560
Leafy vegetables	180	90–350
Legumes	100	80–130
Root vegetables	340	90–440
Garden fruits	110	10–250

Sources: Baes *et al.*, 1984; Hoffman and Baes, 1979; Shor *et al.*, 1982.

Farm animals are exposed to facility-emitted pollutant via ingestion of contaminated forage, grains, and soil. Ingestion of contaminated soil can be a significant exposure pathway for terrestrial organisms, because lipophilic compounds tend to accumulate in soil. Travis and Hattemer-Frey (1987) showed that the ingestion of contaminated soil accounted for 20% of the total daily intake of TCDD by cattle.

The predicted daily intake of pollutants by agricultural animals can be estimated by multiplying the concentration of pollutant in forage, grains, and soil by the quantity of forage, grains, and soil consumed that originated from the study area (i.e., the area affected by facility emissions) and the appropriate chemical-specific, species-specific BTF. Pollutant concentration in a given animal product (meat or milk) on a whole-tissue basis is calculated according to the following equation:

$$C_t = [(Q_f \times F_f \times C_f) + (Q_g \times F_g \times C_g) + (Q_s \times C_s)] \times B_a \qquad (14)$$

where:
C_t = pollutant concentration in a product ($\mu g/g$)
Q_f = quantity of forage consumed by the animal (kg DW/day)
F_f = fraction of forage consumed by the animal that originates from the contaminated area
C_f = pollutant concentration in forage ($\mu g/g$ DW)
Q_g = quantity of grain consumed by the animal (kg DW/day)
F_g = fraction of grain consumed by the animal that originates from the contaminated area
C_g = pollutant concentration in grain ($\mu g/g$ DW)
Q_s = quantity of soil consumed by the animal (kg DW/day)
C_s = pollutant concentration in soil ($\mu g/g$ DW)
B_a = biotransfer factor for the animal product group (day/kg)

Beef cattle consume about 2.66 kg DW of forage and 5.0 kg DW of grains per day, while dairy cattle consume 11.0 and 7.1 kg DW of forage and grains per day, respectively (Shor et al., 1982). It was assumed that 100% of the forage and 60% of the grain consumed by agricultural animals is locally grown (i.e., F_f equals 1.0 and F_g equals 0.6).

Fries (1987) concluded that since lactating dairy cattle are rarely pastured, it is unlikely that soil ingestion would exceed 1% of dry matter intake. Although beef cattle do spend some time on pasture, they are usually confined to feed lots (Fries, 1987). Hence, we assumed that soil ingestion accounts for 1% of dry matter intake for both dairy cattle (0.18 kg DW/day) and beef cattle (0.08 kg DW/day).

Sheep ingest 1.6 kg DW of forage per day (no grains) (Shor et al., 1982), while chickens and pigs consume 0.07 and 3.5 kg DW of grains per day (no forage) (Ng et al., 1982). Fries (1987) reported that sheep will ingest slightly more soil than cattle as represented by the percentage of dry matter intake, while chickens and pigs, which are rarely pastured, would ingest similar amounts of soil as sheep. Hence, we assumed that soil ingestion equals 2% of dry matter intake for sheep (0.03 kg DW per day) and 1% of dry matter intake for chickens and pigs (0.001 and 0.035 kg DW per day, respectively).

ESTIMATING THE DAILY INTAKE OF POLLUTANTS THROUGH THE FOOD CHAIN

For risk assessment purposes, an important objective in evaluating the environmental fate of facility-emitted pollutants is predicting the pathways of human exposure. Three routes exist by which humans can be exposed to pollutants emitted from an MWC through the terrestrial food chain: (1) direct ingestion of contaminated soil, (2) consumption of contaminated plants, and (3) consumption of contaminated animal products (i.e., meats and dairy products).

Table 3. Average daily U.S. human consumption values

Food group	Intake (g DW/day)
Milk and dairy products	40.17
Beef	33.73
Beef liver	6.30
Poultry	12.05
Lamb	1.64
Pork	10.86
Eggs	6.99
Potatoes	30.06
Leafy vegetables	2.59
Legumes (dried)	3.11
Legumes (nondried)	0.78
Root vegetables	3.34
Garden fruits	10.84
Grains and cereals	77.60

Source: Adapted from wet-weight consumption rates reported in Yang and Nelson (1986) and the conversion factors listed in Table 7 (Baes *et al.*, 1984).

Daily intake from the ingestion of contaminated food items is estimated by multiplying the concentration of pollutant in animal products, dairy products, plants, and soil by the average daily human consumption value (g DW/day) (Table 3) and by the fraction of locally grown food consumed (Table 4). Hence, consumption of each food group is weighted by the fraction of that food group that originates from the study area.

Direct Ingestion of Contaminated Soil

Humans can be exposed to pollutants by directly eating contaminated soil or by inadvertent hand-to-mouth transfer of soil. The TFC model calculates human daily pollutant intake from ingesting contaminated soil by:

$$I_s = C_s \times Q_s \times E \tag{15}$$

where: I_s = pollutant intake due to direct ingestion of contaminated soil (μg/day)

C_s = concentration of pollutant in untilled soil after the total period of deposition (μg/g DW)

Q_s = quantity of soil ingested by adults (g DW/day)

E = exposure duration adjustment (unitless)

The intentional ingestion of nonfood items, or pica behavior, occurs mostly in children from ages one to six years. The exposure duration adjustment factor is used to more effectively distribute this exposure over the entire lifetime of the individual. When modeling carcinogenic pollutants, an exposure duration adjustment factor should be used. E should be set equal to 1.0 for all noncarcinogens and for all carcinogens when determining the total daily intake of pollutants by adults. When estimating daily pollutant intake by children from

Table 4. Average percentage of locally produced food items

Food group	All households	Urban households	Rural households (nonfarm)	Rural households (farm)
All meat except poultry	5.25	0.78	6.14	44.2
Poultry and fish	7.38	2.85	11.9	34.3
Dairy products	4.00	0.00	3.99	39.9
Eggs	7.51	0.62	9.14	47.9
Potatoes	9.30	1.21	14.6	44.8
Vegetables	21.2	7.55	35.7	70.0
Legumes (dried)	4.44	2.63	6.90	16.7
Legumes (nondried)	21.2	7.55	35.7	70.0
Root vegetables	21.2	7.55	35.7	70.0
Garden fruits	8.11	3.24	13.2	31.3
Grains and cereals	0.43	0.13	0.82	1.56

Source: U.S. Department of Agriculture (1966).

ingestion of contaminated soil, E is calculated by dividing the duration of exposure (e.g., 5 years) by the assumed lifetime of the child (e.g., 70 years).

Consumption of Contaminated Plants

Humans can also be exposed to pollutants emitted from MWCs by consuming contaminated plants. The total amount of pollutant ingested daily as a result of this pathway is calculated using the following equation:

$$I_p = C_p \times F_p \times E_p \tag{16}$$

where: I_p = daily pollutant intake due to the consumption of contaminated plants (μg/day)
C_p = concentration of pollutant in the plant group (μg/g DW)
F_p = fraction of the plant group that originates from the contaminated area (unitless) (Table 4)
E_p = adult daily dietary consumption of the plant group (g DW/day)

Plants imported from outside the affected area are assumed to be uncontaminated. Total daily intake of pollutant via the ingestion of contaminated plants is calculated by summing the contribution of each individual plant group modeled.

Consumption of Contaminated Animal Products

The total amount of pollutant intake by humans from the consumption of contaminated animal products is given by:

$$I_a = C_t \times F_a \times C_h \tag{17}$$

where: I_a = pollutant intake from the consumption of a contaminated animal product (μg/day)
C_t = concentration of pollutant in the animal product (μg/g DW)
F_a = fraction of the animal product that is locally produced (Table 4)
C_h = daily human consumption of the animal product (g DW/day)

Total daily intake of pollutant via consumption of animal products is calculated by summing the contribution of each individual product group modeled.

Total Daily Intake of Pollutant

Total daily intake of pollutant by humans is calculated by:

$$I_t = I_i + I_s + \sum_{i=1}^{np} I_{pi} + \sum_{i=1}^{na} I_{ai} \tag{18}$$

where: I_t = total daily intake of pollutant by humans (μg/day)

I_i = total daily intake of pollutant resulting from inhalation, [concentration of pollutant in air ($\mu g/m^3$) times average adult inhalation rate of 20 m^3/day)] (μg/day)

I_s = total daily intake resulting from the ingestion of contaminated soil (μg/day)

I_{pi} = total daily intake resulting from the consumption of the ith contaminated plant group (μg/day)

np = total number of plant groups

I_{ai} = total daily intake resulting from the consumption of the ith contaminated animal product (μg/day)

na = total number of animal products

The TFC model further assumes that the concentration of pollutant on vegetation consumed by humans is not decreased by food preparation and washing activities, and that 100% of all pollutants ingested are absorbed through the gut. Although gastrointestinal (GI) absorption factors are widely used in health risk assessments, we believe that using GI factors when estimating exposure is incorrect. Risk estimates are obtained by multiplying exposure by potency. Potency values are derived from laboratory experiments in which test animals are given a specific dose of a carcinogen by one of several routes. In these experiments, potency values are based upon *administered* dose, not effective dose to the target product. Risk estimates should also reflect administered dose. Hence, reducing exposure estimates by applying GI factors is not appropriate.

APPLICATION OF THE TERRESTRIAL FOOD CHAIN MODEL: TCDD

TCDD is a highly lipophilic, extremely persistent compound that sorbs strongly to soil and sediment and bioaccumulates in the food chain. Because TCDD is the most potent chemical ever evaluated by EPA (EPA, 1985b), there is widespread fear that exposure to small amounts of TCDD could lead to serious adverse human health effects. Similarly, because combustion is widely recognized as a source of the highly toxic TCDD (Olie *et al.*, 1982; Lao *et al.*, 1983; Tiernan *et al.*, 1983; Hutzinger *et al.*, 1985; Weerasinge and Gross, 1985; Marklund *et al.*, 1986; Rappe *et al.*, 1987), many people living near MWCs are concerned that they will be exposed to high levels of dioxin and subsequently develop cancer.

Model Inputs and Results for TCDD

The TFC model was used to quantify the magnitude of human exposure to TCDD emitted from a typical, modern MWC in the U.S. based on the following inputs. Table 5 lists the predicted maximum emission rates, air concentrations, and deposition rates of polychlorinated dibenzo-p-dioxins (PCDDs or dioxins) and polychlorinated dibenzofurans (PCDFs or furans). These values are expressed in toxic equivalent factors (TEFs) for 12 proposed MWCs (Hattemer-Frey and Travis, 1989). (The toxicity of different PCDD and PCDF isomers is typically expressed in terms of TEFs, which relate the toxicity of all PCDD and PCDF compounds to the known toxicity of TCDD using one of several weighing schemes. Hence, a mixture of PCDDs and PCDFs expressed in TEFs emitted at the rate of 1 g/s is assumed to have the same toxic effect as a release of 1 g of TCDD per s. Since the values presented in Table 5 were taken directly from the risk assessments, weighing schemes used to estimate PCDD and PCDF concentrations in TEFs may not be consistent.) The deposition rate and maximum air concentration for a typical facility ($1.43 \times 10^{-3} \mu g/m^2$/year and $1.09 \times 10^{-8} \mu g/m^3$, respectively) were used for TCDD in this assessment.

Since the purpose of this exercise is to estimate the magnitude of human exposure to TCDD emitted from a typical MWC in the U.S., we entered generic or mean values into the TFC model for all site-specific input parameters. When the TFC is applied to site-specific cases, site-specific versus generic input data should be used, and the user should anticipate that the major pathways and the extent of human exposure may vary.

Table 5. Predicted maximum emission rates, air concentrations, and deposition rates of PCDDs and PCDFs (expressed in TEFs) for 12 proposed MWCs in the U.S.

Facility	Maximum annual average emission rate (ng/sec)[a]	Maximum air concentration (fg/m³)[b]	Maximum deposition rate (ng/m²/year)[c]
Irwindale	640	2.1	1.32
San Diego	1700	48.0	9.84
Los Angeles	19	2.4	0.18
Philadelphia	390	17.0	0.70
Brooklyn	638	20.3	8.93
Stanislaus	1700	44.0	2.77
N. Hempstead	130	8.1	0.31
Milliken	160	6.1	0.31
Boston	470	15.0	1.98
Bloomington	80	20.0	3.72
Pomona	110	40.0	4.28
Montgomery County	376	1.3	0.60
Typical facility[d]	284	10.9	1.43

[a]Nanograms per second.
[b]Femtograms per cubic meter.
[c]Nanograms per square meter per year.
[d]Geometric mean of the 12 facilities evaluated.

Uptake of TCDD by Plants. Since TCDD is extremely lipophilic, and root uptake is inversely correlated with lipophilicity (Briggs et al., 1982), root uptake was not expected to be a major source of vegetative contamination for TCDD. Using a soil concentration of 8.0×10^{-8} μg/g, a soil-to-plant BCF of 4.25×10^{-3} for vegetation consumed by animals and humans (Travis and Arms, 1988), a soil loss constant of 0.06 year^{-1} (Miller and Hoffman, 1983), and equation 2, the concentration of TCDD in vegetation due to root uptake is estimated to be 0.34 femtograms/g (fg/g) DW.

Since 30% of atmospheric TCDD was assumed to exist in the vapor phase (Bidleman, 1988), air-to-leaf-transfer was suspected to be a major pathway of vegetative contamination. Using a log K_{ow} of 6.85 for TCDD (Travis and Hattemer-Frey, 1987), a Henry's Law Constant of 3.6×10^{-3} (EPA, 1986c), an air-to-leaf BCF of 9838 for vegetation consumed by animals and humans was calculated using equation 5 and was used in this analysis. Applying equation 8, the concentration of TCDD on forage due to foliar uptake is estimated to be 43.5 fg/g DW, while the weighted average concentration on plants consumed by humans is estimated to be 3.2 fg/g DW. Again, it was assumed that grains were not contaminated via this pathway.

Deposition of TCDD onto outer plant surfaces was expected to contribute substantially to vegetative contamination, since it was assumed that all of the TCDD emitted from an MWC is sorbed onto particulates. Furthermore, Isensee and Jones (1971) demonstrated that 94% and 63% of the TCDD applied to soybeans and oats, respectively, remained on the leaves for 21 days, indicating that most of the dioxin deposited from the atmosphere onto outer plant surfaces is likely to persist until the plant is harvested or consumed by animals.

Using: an atmospheric deposition rate of 1.4×10^{-3} $\mu g/m^2$/year;
a weathering constant (K_e) of 14 days (18.1 year^{-1}) (Baes et al., 1984);
an intercept fraction (r) of 0.47 [Hoffman and Baes, 1979 (Table 1)];
0.12 years as the length of the growing season (Baes et al., 1984);
a crop yield of 310 g/m^2 [Hoffman and Baes, 1979 (Table 2)]; and
equation 10,

the concentration of TCDD on forage due to deposition was estimated to be 106.5 fg/g DW. Because grains have a protective outer husk, it was assumed that they were not contaminated via deposition of facility-emitted pollutants.

Using the plant-specific intercept fractions and crop yields for fruits and vegetables listed in Tables 1 and 2, respectively, and 0.16 years as the length of the growing season (Baes et al., 1984), the weighted-average concentration of TCDD on exposed plants consumed by humans due to deposition equals 2.6 fg/g DW (Table 6). The percentage of TCDD contamination resulting from the direct deposition of facility-emitted pollutants onto outer plant parts for individual food crops consumed by humans is listed in Table 6.

Thus, the total concentration of TCDD on forage equals 150 fg/g DW, 71% of which was due to deposition, 0.2% to root uptake, and 29% to foliar uptake (Table 6). Forage was about 500 times more contaminated with TCDD due to facility emissions than were grains (150 fg/g versus 0.3 fg/g). The predicted concentration of TCDD (weighted average) on plants consumed by humans was 6.1 fg/g DW, 42% of which was due to deposition; 53% to air-to-leaf transfer; and 6% to root uptake. Table 6 gives the percentage of vegetative contamination due to deposition, air-to-leaf transfer, and root uptake for individual plant groups consumed by humans.

Uptake of TCDD by Animals. Because of its low water solubility and high lipophilicity, TCDD readily accumulates in living organisms. To estimate the amount of TCDD likely to accumulate in animal products, BTFs for all animal groups were determined. Washburn (1989) used the elimination rate constants reported by Jensen and Hummel (1982) and

Table 6. Percentage of vegetative contamination due to deposition, foliar uptake, and root uptake of TCDD

	TCDD concentration (fg/g)[a]	Percentage due to deposition	Percentage due to foliar uptake	Percentage due to root uptake
Crops consumed by humans				
Potatoes	0.3	0	0	100
Leafy vegetables	110.0	60	39	0.3
Legumes	49.1	12	88	0.7
Root vegetables	0.3	0	0	100
Fruits	76.8	45	54	0.4
Grain	0.3	0	0	100
Weighted-average concentration crops consumed by animals				
Forage	150.0	71	29	0.2
Grain	0.3	0	0	100

[a]Femtograms per gram dry weight basis.

Jensen *et al.* (1981) to calculate steady-state BTFs for cow milk and beef. These steady-state BTFs (0.03 day/kg for milk and 0.8 day/kg for beef) were converted to a DW basis using wet-to-dry weight conversion factors of 0.13 and 0.385 for cow milk and beef, respectively (Table 7) (Baes *et al.*, 1984). Thus, BTFs of 2.08 and 0.23 day/kg DW were used to predict the accumulation of TCDD in cow beef and milk.

Data on the bioaccumulation of TCDD in other agricultural livestock, including pigs, chickens, and sheep, are scarce. As a result, we estimated the BTFs for chicken, lamb, pork,

Table 7. Dry-to-wet weight conversion

Food group	Factor
Beef	0.385
Beef liver	0.303
Dairy products	0.130
Pork, chicken, and lamb	0.400
Eggs	0.260
Leafy vegetables	0.066
Exposed produce	0.126
Protected produce	0.222
Forage and grains	0.888

Sources: Baes *et al.*, 1984; Watt and Merrill, 1963.

and eggs that were required to yield the measured concentration of TCDD in poultry, eggs, lamb, and pork reported by Beck *et al.* (1989) and converted those values to a DW basis using the conversion factors in Table 7. This approach yielded DW BTFs of 0.02, 0.56, 119, and 90 kg DW/day for lamb, pork, poultry, and eggs, respectively. Since data on the measured concentration of TCDD in beef liver were not available, the BTF used for beef was also applied to beef liver.

Table 8 gives the predicted TCDD concentration in animal products resulting from the ingestion of contaminated soil, grains, and forage. While ingestion of contaminated forage accounted for most of the total TCDD concentration in beef, beef liver, dairy products, and lamb (up to 85%), soil ingestion accounted for 99% of the total TCDD concentration in chicken, eggs, and pork, and for 21% and 18% of the total concentration in beef/beef liver and dairy products, respectively. These data confirm that ignoring soil ingestion as a source of exposure by agricultural animals will lead to an underestimation of the concentration of pollutant in animal products, especially for lipophilic compounds. Ingestion of grains contaminated from facility emissions was not a major pathway of TCDD exposure by agricultural animals. Table 8 lists the predicted concentration of TCDD in animal products.

Table 8. Predicted TCDD concentration in animal products from ingesting contaminated soil, forage, and grains

Product group	Total concentration (fg/g DW)[a]	Percentage due to ingesting forage	Percentage due to ingesting grains	Percentage due to ingesting soil
Beef/beef liver	1090	76	0.2	24
Dairy products	449	85	0.1	15
Pork	32	0	1	99
Chicken	193	0	1	99
Lamb	6	82	0	18
Eggs	146	0	1	99

[a]Femtograms per gram dry weight basis.

Total Daily Intake of TCDD by Humans. Table 9 gives predicted average daily intake of TCDD by humans from the ingestion of contaminated food items and inhalation of facility-emitted TCDD. A point estimate of the total daily intake of TCDD from inhalation is included for comparison purposes. These data show that the food chain, especially meat and dairy products, accounts for about 93% of human exposure to TCDD emitted from a typical, modern MWC. Consumption of contaminated vegetation, eggs, and soil is not a major pathway of human exposure. The average daily intake of TCDD from both direct and indirect pathways of human exposure is estimated to be 3.7 picograms/day (pg/day), which corresponds to a cancer risk level of 8×10^{-6}.

Table 9. Predicted average daily intake of TCDD by individuals living near a typical MWC

	Daily intake (pg/day)[a]	Percent of total daily intake
Fruits and vegetables	*0.14*	*3.9*
Potatoes	0.001	0.03
Leafy vegetables	0.06	1.7
Legumes (nondried)	0.008	0.22
Legumes (dried)	0.007	0.18
Root vegetables	0.0002	0.01
Garden fruits	0.067	1.8
Grains and cereals		
Milk and dairy products	*0.72*	*19.7*
Meats (total)	*2.43*	*65.9*
Beef	1.95	53.0
Beef liver	0.37	10.0
Pork	0.018	0.5
Poultry	0.086	2.4
Lamb	0.001	0.01
Eggs	*0.08*	*2.1*
Soil ingestion[b]	*0.06*	*1.6*
Inhalation[c]	*0.25*	*6.8*
Total	*3.68*	*100*

[a]Picograms per day.
[b]Calculated using a soil concentration of 1.6×10^{-6} micrograms per gram dry weight basis.
[c]Calculated using an average adult inhalation rate of 20 cubic meters per day.

VALIDATION OF THE TERRESTRIAL FOOD CHAIN MODEL

To validate the TFC model, input parameters for TCDD were modified to simulate background environmental conditions. Then model predictions of the background daily intake of TCDD by humans and concentrations of TCDD in various media were compared to actual measured data. To estimate background levels of TCDD using the TFC model, the measured background concentration of TCDD in urban air [2×10^{-8} $\mu g/m^3$ (Rappe and Kjeller, 1987)] was used in the TFC model. Next, the measured background air concentration (2.0×10^{-8} $\mu g/m^3$) was multiplied by the percentage of TCDD in the particulate phase (0.7) and the geometric mean maximum deposition velocity estimated for a typical, modern MWC, 6.3×10^4 m/year (Travis and Hattemer-Frey, 1989) to derive a "background" deposition rate of 8.8×10^{-4} $\mu g/m^2$/year. Using the background deposition rate and air concentration values and assuming that all food consumed originated from the contaminated area, TFC model predictions of the concentration of TCDD in various environmental media agree well with

Table 10. Comparison of background concentrations of TCDD predicted by the TFC model with measured background environmental concentrations

Media	Concentration predicted by TFC (pg/g WW)[a]	Background concentration (pg/g WW)	Reference for background data
Beef	0.33	0.15	Beck *et al.*, 1989
Beef liver	0.26	NA[b]	—
Dairy products	0.05	0.007	Beck *et al.*, 1987
		0.009	Startin *et al.*, 1989
		0.015	Fürst *et al.*, 1989
Pork	0.008	0.008	Beck *et al.*, 1989
Poultry	0.048	0.045	Beck *et al.*, 1989
Lamb	0.002	0.003	Beck *et al.*, 1989
Eggs	0.023	0.02	Beck *et al.*, 1989
Leafy vegetables	0.007	0.005	Beck *et al.*, 1989
Garden fruits	0.018	0.006	Travis and Hattemer-Frey, 1989
Forage	0.11	0.08	Travis and Hattemer-Frey, 1989
Grain	0.0002	0.006	Travis and Hattemer-Frey, 1989
Soil	0.97	0.96	Nestrick *et al.*, 1988
		1.4	Rappe and Kjeller, 1987

[a]Picograms per gram wet-weight or whole-tissue basis. TFC model predictions were converted from a dry- to a wet-weight basis using the conversion factors listed in Table 7.
[b]Measurement not available.

measured background concentrations (Table 10). The total background daily intake of TCDD through the terrestrial food chain obtained from the TFC model (53 pg/day) agrees reasonably well with estimates of 39 and 16 pg/day reported by Travis and Hattemer-Frey (1987) and Beck *et al.* (1989) and a pharmacokinetically derived estimate of 29 pg/day (Travis and Hattemer-Frey, 1989). These results demonstrate that the TFC is a valid model for quantifying total daily intake of pollutants emitted from an MWC.

APPLICATION OF THE TERRESTRIAL FOOD CHAIN MODEL TO CADMIUM

Ingestion is the primary route by which humans are exposed to metals present in the environment (Belcher and Travis, 1989). Metals are most notably different from organic compounds in that they do not decay or degrade in the environment. Unlike organics, metals can enter plant tissues only through root uptake and direct deposition.

Model Inputs and Results

The TFC model was used to quantify the magnitude of human exposure to cadmium emitted from a typical MWC in the U.S. based on the following inputs: a deposition rate of 7.3 $\mu g/m^2$/year and a maximum air concentration of 4.0×10^{-6} $\mu g/m^3$ (Belcher and Travis, 1989). Again, since the purpose of this exercise was to estimate the magnitude of human exposure to cadmium emitted from a typical MWC, we entered generic or mean values for all site-specific input parameters.

Uptake of Cadmium by Plants. Since root uptake is the primary pathway of transport of many metals to plants (Allaway, 1968), this pathway was expected to be the primary route of plant uptake of cadmium. Using a soil-to-plant BCF of 0.39 for forage, 0.05 for grains, 0.09 for potatoes, 1.18 for leafy vegetables, 0.24 for fresh and dried legumes, 1.98 for root vegetables, and 1.16 for garden fruits (Belcher and Travis, 1989; Heffron, 1980; EPA, 1985a) and assuming that cadmium does not degrade in soil (i.e., K_s equals zero), the concentration of cadmium on forage, grains, and food crops consumed by humans (a weighted average) due to root uptake equals 0.39 nanograms/g (ng/g), 0.05 ng/g, and 0.15 ng/g, respectively.

Using a weathering constant (K_e) of 31.88 year^{-1} (Baes et al., 1984; EPA, 1985a), an intercept fraction (r) of 0.47 (Hoffman and Baes, 1979) (Table 1), 0.12 years as the length of the growing season (Baes et al., 1984), and a crop yield of 310 g/m^2 (Hoffman and Baes, 1979) (Table 2), the concentration of cadmium on forage due to deposition was estimated to be 0.34 ng/g DW. The weighted-average concentration of TCDD on exposed vegetables and fruits due to deposition is estimated to equal 0.01 ng/g DW. This value was obtained by using the plant-specific intercept fractions and crop yields for fruits and vegetables listed in Tables 1 and 2, respectively, and 0.16 years as the length of the growing season (Baes et al., 1984).

Summing the appropriate figures listed above yields a total concentration of cadmium on forage of 0.73 ng/g DW, 53% of which was due to root uptake and 47% to deposition (Table 11). In this case, forage was only about 15 times more contaminated with cadmium due to facility emissions than were grains (0.7 ng/g versus 0.05 ng/g). The weighted-average concentration of TCDD on fruits and vegetables consumed by humans was 0.16 ng/g DW, 95% of which is due to root uptake (Table 11). Garden fruits and leafy vegetables accumulated the highest cadmium levels of all plant groups consumed by humans.

Uptake of Cadmium by Animals. The following BTFs (DW basis) were incorporated into the model to predict the bioaccumulation of cadmium in animal products (all units are day/kg): beef (0.003), beef liver (0.135), lamb (0.03), pork (0.03), poultry (12.45), dairy products (0.006), and eggs (0.65) (Belcher and Travis, 1989). The predicted concentration of cadmium in animal products resulting from the ingestion of facility-contaminated soil, grains, and forage is given in Table 12. These data show that ingestion of contaminated soil was the major pathway of exposure for pigs and chickens, while ingestion of contaminated forage was the primary route of exposure for beef cattle, dairy cattle, and sheep. Chickens accumulated the highest concentration of cadmium due to facility emissions.

Total Daily Intake of Cadmium by Humans. Table 13 gives the predicted average daily intake of cadmium by individuals living near a typical MWC from ingesting contaminated food items, incidental soil ingestion, and inhalation. These data show that the food chain accounts for approximately 98% of human exposure to cadmium. Unlike TCDD, the primary pathway of

Table 11. Percentage of vegetative contamination due to deposition,
foliar uptake, and root uptake for cadmium

	Total cadmium concentration $(ng/g\ DW)^a$	Percentage due to deposition	Percentage due to foliar uptake	Percentage due to root uptake
Crops consumed by humans				
Potatoes	0.09	0	0	100
Leafy vegetables	1.4	14	0	86
Legumes	0.26	7	0	93
Root vegetables	2.0	0	0	100
Garden fruits	1.3	8	0	92
Grains	0.05	0	0	100
Weighted average	0.16	5	0	95
Crops consumed by animals				
Forage	0.73	47	0	53
Grains	0.05	0	0	100

^{a}Nanograms per gram dry weight basis.

human exposure is consumption of contaminated plants, especially root vegetables and garden fruits, which accounted for 73% of total daily intake of cadmium. Consumption of contaminated meat and dairy products accounted for only 11% of total daily intake, and inhalation was a negligible source of human exposure. The average daily intake of cadmium from both direct and indirect pathways of human exposure is estimated to be 4.9 ng/day.

PROBABILISTIC RISK ASSESSMENT

To provide a new perspective on human exposure to TCDD emitted from MWCs, we performed a PRA that takes into account variability in facility design, location, and population distributions around existing MWCs. The population-weighted average, actual maximum, and theoretical maximum individual exposure concentrations $(\mu g/m^3)$ resulting from a unit release rate of 1 g of PCDDs and PCDFs expressed in toxic equivalents (hereafter referred to as TCDD toxic equivalents) per s were calculated for 24 operating facilities across the U.S. Data on facility characteristics were obtained from EPA (1987) and are presented in Table 14.

Exposure concentrations were calculated within a 50-km radius of each facility using the Industrial Source Complex Long-Term (ISCLT) model, a Gaussian plume model developed by EPA (1986a). Although the 50-km radius was arbitrarily selected, actual facility parameters and site-specific meteorological and population data were used in this analysis.

Table 12. Predicted concentration of cadmium in animal products
from ingestion of contaminated soil, forage, and grains

Product group	Total concentration (pg/g DW)[a]	Percentage due to ingesting forage	Percentage due to ingesting grains	Percentage due to ingesting soil
Beef	10.9	53	4	43
Beef liver	489	54	4	43
Dairy products	65.0	68	2	30
Pork	22.5	0	13	87
Chicken	274.0	0	7	93
Lamb	48.7	65	0	35
Eggs	14.3	0	10	90

[a]Picograms per gram dry-weight basis.

The number and location of individuals residing within a 50-km radius of each facility were obtained from 1980 census data.

The population-weighted annual average atmospheric concentration was obtained by dividing the 50-km radius area around each facility into 16 radial sectors and 14 concentric rings yielding 224 sector segments. ISCLT was then applied to predict the annual average air concentration at each of the 224 intersecting points (e.g., G1 and G2 in Figure 1). The number of people residing in the ith sector segment (P_i) was multiplied by C_i, the annual average air concentration predicted by ISCLT at the two points bordering each sector segment (Figure 1). These exposure values were added together and divided by the total population to obtain the population-weighted average exposure concentration. This approach is represented mathematically by the equation:

$$[(C_1 \times P_1) + (C_2 \times P_2) + \quad \ldots \quad + (C_{224} \times P_{224})] \div P_t \qquad (19)$$

where P_t is the total population living within a 50-km radius area of a given MWC.

The maximum air concentration is defined as the maximum ground-level air concentration in any sector segment inhabited by people. Conversely, the theoretical maximum air concentration represents the maximum ground-level air concentration in any of the 224 sector segments regardless of whether any people actually live in that sector segment.

Table 13. Predicted average daily intake of cadmium
by individuals living near a typical MWC

	Daily intake (ng/day)[a]	Percentage of total daily intake
Fruits and vegetables (total)	*3.58*	*72.6*
Potatoes	0.24	5.0
Leafy vegetables	0.75	15.2
Legumes, nondried	0.04	0.9
Legumes, dried	0.03	0.7
Root vegetables	1.38	28.1
Garden fruits	1.1	22.2
Grains	0.04	0.7
Milk and Dairy Products	*0.10*	*2.1*
Meats (total)	*0.42*	*8.7*
Beef	0.02	0.4
Beef liver	0.16	3.2
Pork	0.01	0.3
Poultry	0.23	4.7
Lamb	0.004	0.1
Eggs	*0.01*	*0.2*
Soil Ingestion[b]	*0.74*	*15.0*
Inhalation[c]	*0.08*	*1.6*
Total	*4.93*	*100*

[a]Nanograms per day.
[b]Calculated using a soil concentration of 2.0×10^{-2} micrograms per gram dry-weight basis.
[c]Calculated using an average adult inhalation rate of 20 cubic meters per day.

Table 14. Operating characteristics and location of the 24 MWCs used in the PRA

	Stack height (m)	Stack diameter (m)	Exit gas velocity (m/s)	Exit gas temperature (°K)	Latitude (°N)	Longitude (°W)
Livingston, MT	5	1.8	3.5	420	45.4	110.3
Ogden, UT	46	1.5	13.4	477	41.1	111.6
Collegeville, MN	37	3.6	0.7	350	45.4	94.3
Waukesha, WI	32	2.1	9.2	420	43.0	8.1
Stuttgart, AZ	11	0.8	9.3	420	34.3	91.3
Palestine, TX	15	0.9	5.5	350	31.5	95.4
Pinellas County, FL	49	3.7	9.2	420	27.6	82.5
Lewisburg, TN	15	1.2	6.6	350	35.3	86.5
Baltimore RESCO, MD	96	1.8	13.4	506	39.2	76.4
Euclid, OH	38	0.8	7.3	505	41.3	81.3
Harrisburg, PA	49	0.6	8.3	420	40.2	76.5
Huntington, NY	38	2.1	12.1	494	40.5	73.3
Durham, NH	35	1.8	11.5	455	43.1	70.5
Bellingham, WA	11	1.1	6.2	420	48.5	122.3
Meredith, NH	24	0.9	11.1	830	43.6	71.5
Pittsfield, NH	8	0.6	21.2	420	43.3	71.3
Oswego County, NY	43	0.6	22.1	420	43.3	76.4
Johnsonville, SC	9	2.1	1.8	420	33.8	79.5

	Stack height (m)	Stack diameter (m)	Exit gas velocity (m/s)	Exit gas temperature (°K)	Latitude (°N)	Longitude (°W)
Ames, IA	61	2.4	1.3	505	42.0	93.6
Haverhill, MA	99	2.1	19.0	435	42.7	71.2
Waxahachie, TX	27	0.9	11.8	350	32.4	96.9
Lakeland, FL	61	3.7	1.3	400	28.0	82.0
Sitka, AK	24	1.0	4.0	420	57.1	135.3
Osceola, AR	9	1.2	5.6	420	35.7	89.9
Geometric Mean	*26.5*	*1.4*	*6.6*	*435*		

Source: EPA, 1987.

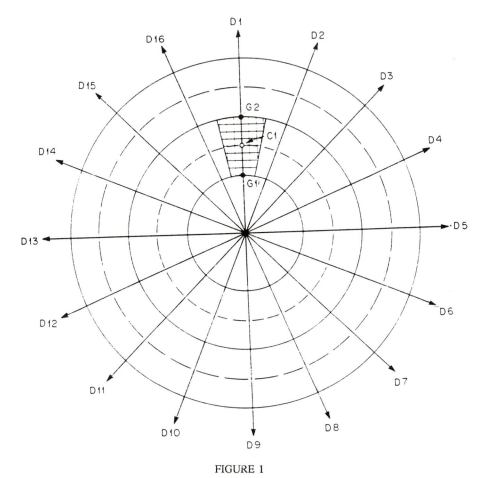

FIGURE 1

Air Dispersion Grid Used to Calculate the Maximum
and Population-Weighted Annual Average Exposure Concentrations

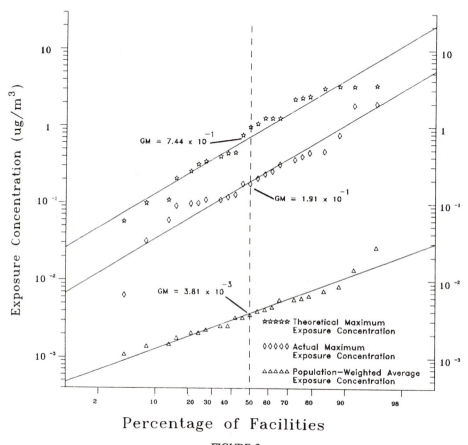

FIGURE 2

Predicted Maximum and Population-Weighted Exposure Concentrations
for TCDD Based on a One Gram Per Second Unit Emission Rate

Figure 2 shows that the maximum and population-weighted annual average ground-level exposure concentrations resulting from a unit release (1 g/s) of TCDD are lognormally distributed. This distribution reflects the variability in individual exposures resulting from the combustion of municipal solid waste (MSW) at a typical facility in the U.S., taking into account differences in incinerator design (i.e., stack height, stack gas temperature, and stack exit velocity), site-specific meteorological conditions, and population density. These data demonstrate that given a unit release of TCDD and assuming reasonable conditions (i.e., using the actual maximum), the maximally exposed individual living near a typical MWC is likely to encounter a risk level 50 times higher than the corresponding risk level for the average exposed individual (Figure 2). Under worst-case conditions (i.e., relying on the theoretical maximum exposure), the maximally exposed individual would experience a mean risk level 200 times greater than the risk level associated with the mean population-weighted average exposure concentration.

The fact that the line depicting the distribution of the population-weighted average exposure concentration associated with a release (1 g/s) of TCDD in Figure 2 is not parallel to the other two lines suggests that the numbers of people living near MWCs in the U.S. is not consistent. The flatness of the population-weighted average exposure distribution relative to the maximum exposure distributions suggests that fewer people are being exposed to MWC emissions than predicted. These results seem to indicate that larger facilities are located in less populated (e.g., rural) areas, while smaller MWCs are generally built in more heavily populated (e.g., urban or metropolitan) areas.

Next, the range of TCDD as well as PCDD and PCDF emissions (expressed in TCDD toxic equivalents) in g/s at 12% oxygen from existing MWCs were obtained from EPA (1987) (Figure 3). Figure 3 shows that emissions of TCDD and TCDD-equivalents are, likewise, lognormally distributed. Not surprisingly, TCDD-equivalent emissions are about one order of magnitude higher than TCDD emissions. The fact that the two distributions are parallel, however, indicates that the ratio of PCDD and PCDF isomers emitted from MWCs across the U.S. are relatively consistent.

The distribution of actual TCDD toxic equivalent emissions (Figure 3) was merged with the exposure concentration distributions based on a unit emission rate (Figure 2) to obtain the probability distribution of *actual* population-weighted annual average and maximum ground-level air concentrations (Figure 4). Again these distributions depict air concentrations that result from the combustion of MSW at a typical facility in the U.S. These actual TCDD toxic equivalent exposure concentrations were multiplied by the average human inhalation rate of 20 m^3/day and divided by human body weight (70 kg) to obtain daily intake via inhalation in mg/kg/day. These intake estimates were then multiplied by the cancer potency factor for TCDD [$(1.56 \times 10^5$ mg/kg/day$)^{-1}$ (EPA, 1985b)] to obtain a distribution of cancer risk associated with inhaling PCDDs and PCDFs emitted from a typical MWC in the U.S. (Figure 5).

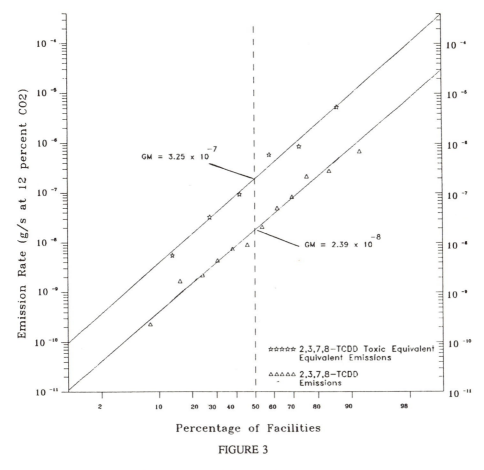

FIGURE 3

Range of 2,3,7,8-TCDD and TCDD-Toxic Equivalent Emissions
From Existing Municipal Waste Combustors (US EPA, 1987)

Figure 5 shows that at 50% of the facilities in the U.S., the average exposed individual would encounter a cancer risk level associated with exposure to facility-emitted PCDDs and PCDFs via inhalation that is 50 times lower than the maximally exposed individual (5.5×10^{-8} versus 2.8×10^{-6}). Since the risk level associated with the theoretical maximum exposure concentration is 4 times higher than the risk level associated with actual maximum exposures, risk assessments that rely on the theoretical maximum exposure concentration to quantify inhalation risks may overestimate the true maximum individual inhalation risk by a factor of four. Furthermore, there is a 10% chance that the maximally exposed individual living near a typical MWC would encounter a lifetime cancer risk of 10^{-4}, while only 10% of the average exposed individuals living near a typical MWC are expected to experience a lifetime cancer risk greater than 10^{-6}. These data suggest that a small percentage (roughly 10%) of operating MWCs in the U.S. may be releasing unacceptably high levels of PCDDs and PCDFs.

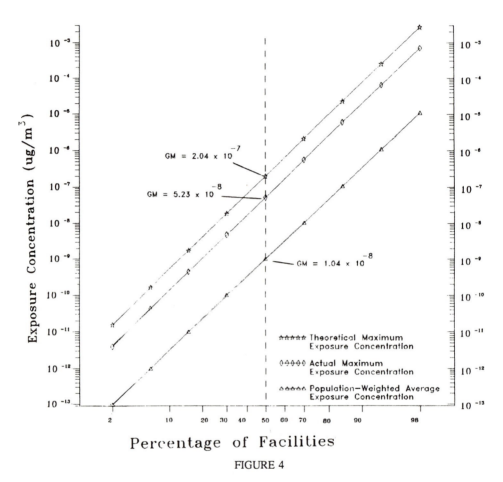

FIGURE 4

Probability Distribution of the Average and Maximum
TCDD-Toxic Equivalent Exposure Concentrations in Air
Near an Arbitrary Municipal Waste Combustor in the U.S.

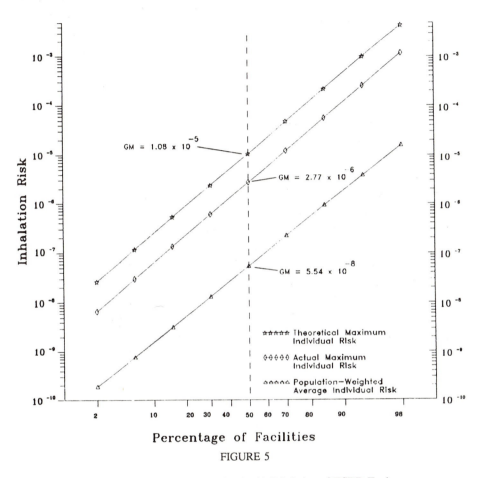

FIGURE 5

Distribution of Risks Associated with Inhalation of TCDD-Toxic
Equivalent Emissions From an Arbitrary Municipal Waste Combustor in the U.S.

CONCLUSIONS

Since organic compounds tend to accumulate in the media in which they are most soluble, assessing the magnitude of exposure to pollutants emitted from an MWC depends largely on being able to predict their bioaccumulation in the food chain. The TFC model, a multimedia transport model, was used to assess the magnitude of human exposure to TCDD and cadmium resulting from the combustion of MSW. The model quantifies the amount of pollutant entering the terrestrial food chain and the extent of human exposure from inhalation and ingestion of contaminated agricultural produce, meats, milk, eggs, and soil. Results show that the food chain, especially consumption of meat and dairy products, accounts for 93% of human exposure to TCDD emitted from a typical MWC. The average daily intake of TCDD due to facility emissions is estimated to be 3.7 pg/day, which corresponds to an individual lifetime risk level of 8×10^{-6}. Similarly, the food chain accounted for about 98% of total daily intake of cadmium, which was estimated to be 4.9 ng/day. Unlike TCDD, however, consumption of contaminated produce was the major pathway of human exposure to facility-emitted cadmium.

To provide a new perspective on human exposure to pollutants emitted from a typical U.S. MWC, we performed a PRA of inhalation exposures. The distributions generated from this analysis reflect the variability in human exposure via inhalation resulting from the combustion of MSW at a typical MWC in the U.S. taking into account differences in facility design and location. Our results show that the average individual living near a typical MWC is likely to encounter an inhalation risk of 5.5×10^{-8} and that only 10% of the average-exposed individuals living near a typical MWC are expected to experience a risk level greater than 10^{-6}. We recommend that future risk assessments report the population-weighted average risk level along with the (actual) maximum risk level to provide more detailed information about the range of risks likely to occur at a given facility.

ACKNOWLEDGMENTS

Research was sponsored by EPA under an Interagency Agreement applicable under Martin Marietta Energy Systems, Inc., Contract No. DE-AC05-84OR21400 with the U.S. Department of Energy.

REFERENCES

Allaway, W.H., 1968, Agronomic controls over environmental cycling of trace elements, *Advan. Agron.*, 20:235-273.

Bacci, E., Calamari, D., Gaggi, C., and Vighi, M., 1989, Bioconcentration of organic chemical vapors in plant leaves: Experimental measurements and prediction, *Chemosphere*, in press.

Bacci, E., and Gaggi, C., 1987, Chlorinated hydrocarbon vapors and plant foliage: Kinetics and applications, *Chemosphere*, 16:2515-2522.

Bacci, E., and Gaggi, C., 1986, Chlorinated pesticides and plant foliage: Translocation experiments, *Bull. Environ. Contam. Toxicol.*, 37(6):850-857.

Bacci, E., and Gaggi, C., 1985, Polychlorinated biphenyls in plant foliage: Translocation or volatilization from contaminated soils, *Bull. Environ. Contam. Toxicol.*, 35(5):673-685.

Baes, C. F., III, 1982, Prediction of radionuclide K_d values from soil-plant concentration ratios, *Trans. Amer. Nuc. Soc.*, 41:53-54.

Baes, C. F., III, Sharp, R. D., Sjoreen, A., and Shor, R., 1984, "A Review and Analysis of Parameters for Assessing Transport of Environmentally Released Radionuclides Through Agriculture," ORNL-5786, U.S. Department of Energy, Oak Ridge National Laboratory, Oak Ridge, Tennessee.

Beall, M. L., Jr., and Nash, R. G., 1971, Organochlorine insecticide residues in soybean plant tops: Root versus vapor sorption, *Agron. J.*, 63:460-464.

Beck, H., Eckart, K., Mathar, W., and Wittowski, R., 1989, PCDD and PCDF body burden from food intake in the Federal Republic of Germany, *Chemosphere*, 18(1-6):417-422.

Beck, H., Eckert, K., Kellert, M., Mathar, W., Ruhl, Ch.-S., and Wittowski, R., 1987, Levels of PCDDs and PCDFs in samples of human origin and food in the Federal Republic of Germany, *Chemosphere*, 16:1977-1982.

Belcher, G. D., and Travis, C. C., 1989. An uncertainty analysis of food chain exposures to pollutants emitted from municipal waste combustors, in *Municipal Waste Incineration and Human Health*, C.C. Travis and H.A. Hattemer-Frey, eds., CRC Press, Boca Raton, Florida.

Bidleman, T. F., 1988, Atmospheric processes, *Environ. Sci. Technol.*, 22:361-367.

Briggs, G. G., Bromilow, R. H., and Evans, A. A., 1982, Relationships between lipophilicity and root uptake and translocation of non-ionized chemicals by barley, *Toxicol. Environ. Chem.*, 7:173-189.

Buckley, E. H., 1982, Accumulation of airborne polychlorinated biphenyls in foliage, *Science*, 216:520-522.

Cupitt, L. T., 1980, "Fate of Toxic and Hazardous Materials in the Air Environment," EPA 600/S3-80-084, U.S. Environmental Protection Agency, Washington, D.C.

de Temmerman, L., Vandeputte, R., and Guns, M., 1986, Biological monitoring and accumulation of airborne mercury in vegetables, *Environ. Poll.*, 41:139-151.

Fries, G. F., 1987, Assessment of potential residues in food derived from animals exposed to TCDD contaminated soil, *Chemosphere*, 16(8/9):2123-2128.

Fürst, P., Fürst, C., and Groebel, W., 1989, "Levels of PCDDs and PCDFs in Food-Stuffs from the Federal Republic of Germany," presented at the Ninth International Symposium on Chlorinated Dioxins and Related Compounds, September 20-23, Toronto, Canada.

Halfon, E., 1985, Regression method in ecotoxicology: A better formulation using the geometric mean functional regression, *Environ. Sci. Technol.*, 19:747-749.

Hattemer-Frey, H. A., and Travis, C. C., 1989a, Characterizing the extent of human exposure to PCDDs and PCDFs emitted from municipal waste combustors, in *Municipal Waste Incineration and Human Health*, C.C. Travis and H.A. Hattemer-Frey, eds., CRC Press, Boca Raton, Florida.

Hattemer-Frey, H. A., and Travis, C. C., 1989b, Pentachlorophenol: Environmental partitioning and human exposure, *Arch. Environ. Contam. Toxicol.*, 18:482-487.

Heffron, C. L., Reid, J. L., Elfving, D. C., Stoewsanson, G. S., Haschek, W. M., Telford, J. N., Furr, A. K., Parkinson, T. F., Bache, C. A., and Gutenmann, W. H., 1980, Cadmium and zinc in growing sheep fed silage corn grown on municipal sludge amended soil, *J. Agric. Food Chem.*, 28:58-61.

Hoffman, F. O., and Baes, C. F., 1979, "A Statistical Analysis of Selected Parameters for Predicting Food Chain Transport and Internal Dose of Radionuclides," ORNL/NUREG/TM-882, Oak Ridge National Laboratory, Oak Ridge, TN.

Hutzinger, O., Blumich, M. J., van den Berg, M. and Olie, K., 1985, Sources and fate of PCDDs and PCDFs: An overview, *Chemosphere*, 14(6/7):581-600.

Isensee, A. R., and Jones, G. E., 1971, Absorption and translocation of root and foliage applied 2,4-dichlorophenol, 2,7-dichlorodibenzo-p-dioxin, and 2,3,7,8-tetrachlorodibenzo-*p*-dioxin, *J. Agric. Food Chem.*, 19(6):1210-1214.

Jensen, D. J., and Hummel, R. A., 1982, Secretion of TCDD in milk and cream following the feeding of TCDD to lactating dairy cows, *Bull. Environ. Contam. Toxicol.*, 29:440-.

Jensen, D. J., Hummel, R. A., Mahle, N. H., Kocher, C. W., and Higgins, H. S., 1981, A residue study on beef cattle consuming 2,3,7,8-TCDD, *J. Agric. Food Chem.*, 29:265-.

Junge, C. E., 1977, Basic considerations about trace constituents in the atmosphere as related to the fate of global pollutants, in *Fate of Pollutants in the Air and Water Environments*, I.H. Suffett, ed., Wiley-Interscience, New York.

Lao, R. C., Thomas, R. S., Chiu, C., Li, K. and Lockwood, J., 1983, Analysis of PCDD-PCDF in environmental samples, in *Chlorinated Dioxins and Dibenzofurans in the Total Environment II*, L. H. Keith, C. Rappe, and G. Choudhary, eds., Butterworth Publishers, Boston.

Lioy, P. L., Harkov, R., Waldman, J. M., Pietarinen, C., and Greenberg, A., 1988, The total human environmental exposure study (THEES) to benzo(a)pyrene: Comparison of the inhalation and food pathways, *Arch. Environ. Health*, 43(4):304-312.

Marklund, S., Kjeller, L-O., Hansson, M., Tysklind, M., Rappe, C., Ryan, C., Collazo, H., and Dougherty, R., 1986, Determination of PCDDs and PCDFs in incineration samples and pyrolytic products, in *Chlorinated Dioxins and Dibenzofurans in Perspective*, C. Rappe, G., Choudhary, and L.H. Keith, eds., Lewis Publishers, Chelsea, Michigan.

Mackay, D., 1982, Correlation of bioconcentration factors, *Environ. Sci. Technol.*, 16:274-278.

Mackay, D., Paterson, S., and Schroeder, W.H., 1986, Model describing the rates of transfer processes of organic chemicals between atmosphere and water, *Environ. Sci. Technol.*, 20(8);810-816.

Miller, C. and Hoffman, F., 1983, An examination of the environmental half-time for radionuclides deposited onto vegetation, *Health Phys.*, 45:731-744.

Mosbaek, H., Tgell, J., and Sevel, T., 1988, Plant uptake of airborne mercury in background areas, *Chemosphere*, 17(6):1227-1236.

Nash, R. G., 1974, Plant uptake of insecticides, fungicides, and fumigants from soil, in *Pesticides in Soil and Water*, W.D. Guenzi, ed., Soil Science Society of America, Inc., Madison, Wisconsin.

Nash, R. G., and Beall, M. L., Jr., 1980, Distribution of silvex, 2,4-D, and TCDD applied to turf in chambers and field plots, *J. Agric. Food Chem.*, 28(3):614-623.

Nestrick, T. J., Lamparski, L. L., Frawley, N. N., Hummel, R. A., Kocher, C. W., Mahle, N. H., McCoy, J. W., Miller, D. L., Peters, T. L., Pillepich, J. L., Smith, W. E., and Tobey, S. W., 1986, Perspectives of a large scale environmental survey for chlorinated dioxins: Overview and soil data, *Chemosphere*, 15(9-12):1453-1460.

Ng, Y., Colsher, C., and Thompson, S., 1982, "Transfer Coefficients for Assessing the Dose from Radionuclides in Meat and Eggs," NUREG/CR-2967, U.S. Government Printing Office, Washington, D.C.

Olie, K., Lustenhouwer, J. W. A., and Hutzinger, O., 1982, Polychlorinated dibenzo-p-dioxins and related compounds in incinerator effluents, in *Chlorinated Dioxins and Related Compounds: Impact on the Environment*, O. Hutzinger, R. W. Frei, E. Merian, and F. Pocchiari, eds., Pergamon Press, Oxford.

Rappe, C., Andersson, R., Bergqvist, P-A., Brohede, C., Hansson, M., Kjeller, L-O., Lindstrom, G., Marklund, S., Nygren, M., Swanson, S. E., Tysklind, M., and Wiberg, K., 1987, Overview on environmental fate of chlorinated dioxins and dibenzofurans: Sources, levels and isomeric pattern in various matrices, *Chemosphere*, 16(8/9):1603-1618.

Rappe, C., and Kjeller, L-O., 1987, PCDDs and PCDFs in environmental samples air, particulates, sediments and soil, *Chemosphere*, 16(8/9):1775-1780.

Seiber, J. N., Kim, Y-H., Wehner, T., and Woodrow, J. E., 1982, "Analysis of Xenobiotics," presented at the Fifth International Congress of Pesticide Chemistry (IUPAC), Kyoto, Japan.

Shor, R. W., Baes, C. F., III, and Sharp, R. D., 1982, "Agricultural Production in the United States by County: A Compilation of Information from the 1975 Census of Agriculture for Use in Terrestrial Food Chain Transport and Assessment Models," ORNL-5768, U.S. Department of Energy, Oak Ridge National Laboratory, Oak Ridge, Tennessee.

Startin, J. R., Rose, M., Wright, C., and Gilbert, J., "Surveillance of British foods for PCDDs and PCDFs," presented at the Ninth Inter. Symp. on Chlorinated Dioxins and Related Compounds, Toronto, Canada, September 20-23, 1989.

Tiernan, T. O, Taylor, M. L., Garrett, J. H., van Ness, G. F., Solch, J. G., Deis, D. A., and Wagel, D. J., 1983, Chlorodibenzodioxins, chlorodibenzofurans and related compounds in the effluents from combustion processes, *Chemosphere*, 12(4/5):595-606.

Travis, C. C., and Arms, A. D., 1988, Bioconcentration of organics in beef, milk, and vegetation, *Environ. Sci. Technol.*, 22(3):271-274.

Travis, C. C., and Arms, A. D., 1987, The food chain as a source of toxics exposure, in *Toxic Chemicals, Health, and the Environment*, L. B. Lave, and A. C. Upton, eds., Plenum Press, New York.

Travis, C. C., and Hattemer-Frey, H. A., 1989, Dioxin: Research needs for risk assessment, *Chemosphere*, in press.

Travis, C. C., and Hattemer-Frey, H. A., 1988, Uptake of organics by aerial plant parts: A call for research, *Chemosphere*, 17(2):277-283.

Travis, C. C., and Hattemer-Frey, H. A., 1987, Human exposure to 2,3,7,8-TCDD, *Chemosphere*, 16(10-12):2331-2342.

Travis, C. C., Holton, G. A., Etnier, E. L., Cook, C., O'Donnell, F. R., Hetrick, D. M., and Dixon, E., 1986, Assessment of inhalation and ingestion population exposures from incinerated hazardous wastes, *Environ. Inter.*, 12:553-540.

U.S. Department of Agriculture (USDA), 1966, "Household food consumption survey, 1965-1966, Report No. 12, food consumption of households in the United States, seasons and year, 1965-1966," U.S. Government Printing Office, Washington, D.C.

U.S. Environmental Protection Agency (EPA), 1987, "Municipal waste Combustion Study: Characterization of the Municipal Waste Combustion Industry," EPA/530-SW-87-021H, Pollutant Assessment Branch, Research Triangle Park, NC.

U.S. Environmental Protection Agency (EPA), 1986a, "Industrial Source Complex (ISC) Dispersion Model User's Guide, Second Edition," EPA 450/4-86-005a, Office of Air Quality Planning and Standards, Research Triangle Park, NC.

U.S. Environmental Protection Agency (EPA), 1986b, "Methodology for the Assessment of Health Risks Associated with Multiple Pathway Exposure to Municipal Waste Combustor Emissions," Office of Air Quality Planning and Standards, Research Triangle Park, NC, and the Environmental Criteria and Assessment Office, Cincinnati, OH.

U.S. Environmental Protection Agency (EPA), 1986c, "Superfund Public Health Evaluation Manual," EPA 540/1-86-060, Office of Emergency Response and Remedial Action, Washington, D.C.

U.S. Environmental Protection Agency (EPA), 1985a, "Environmental Profiles and Hazard Indices for Constituents of Municipal Sludge: Cadmium," Office of Water Regulations and Standards, Washington, D.C.

U.S. Environmental Protection Agency (EPA), 1985b, "Health Assessment Document for Polychlorinated Dibenzo-p-Dioxins," EPA-600/8-84-014F, Washington, D.C.

Washburn, S. T., 1989, The accumulation of chlorinated dibenzo-p-dioxins and dibenzofurans in beef and milk, in *Municipal Waste Incineration and Human Health*, C. C. Travis and H. A. Hattemer-Frey, eds., CRC Press, Boca Raton, Florida.

Watt, B. K., and Merrill, A. L., 1963, "The Composition of Foods," U.S. Department of Agriculture Handbook No. 8, Washington, D.C.

Weerasinghe, N. C. A., and Gross, M. L., 1985, Origins of polychlorodibenzo- p-dioxins (PCDD) and polychlorodibenzofurans (PCDF) in the environment, in *Dioxins in the Environment*, M. A. Kamrin and P. W. Rodgers, eds., Hemisphere Publishing Corporation, Washington, DC.

Yang, Y-Y., and Nelson, C. B., 1986, An estimation of daily food usage factors for assessing radionuclide intakes in the U.S. population, *Health Phys.*, 50(2):245-257.

CURRENT STUDIES ON HUMAN EXPOSURE TO CHEMICALS

WITH EMPHASIS ON THE PLANT ROUTE

Sally Paterson and Donald Mackay

Institute of Environmental Studies
University of Toronto
Toronto, Ontario M5S 1A4

INTRODUCTION

The first step in the assessment of human exposure to an organic contaminant is the estimation of concentrations of the chemical in various environmental media such as air, water, soil and sediment. General multimedia models have been developed which can be used calculate an expected environmental distribution of a chemical based on its physical chemical characteristics.[1,2,3,4]

The next step in the assessment is the estimation of concentrations (e.g. mg/kg) of chemical in the primary human exposure media of air, water and food. Combining concentrations in these exposure media with the human intake rates (e.g. kg/day) by various routes results in the estimation of total dose (e.g. mg/day) to humans. Figure 1 illustrates some major exposure pathways to humans. An example of a comprehensive multimedia model incorporating human exposure is that of McKone and Layton.[4] Estimation of exposure through air inhalation and water ingestion is reasonably straightforward. Concentrations in fish have been well quantified by means of bioconcentration and biomagnification factors. Estimation of chemical levels in grains, vegetables, fruit, dairy products and meat is more difficult. Travis and Arms[5] have correlated reported bioconcentration factors in beef, milk and vegetation with octanol/water partition coefficient (K_{ow}) for a number of chemicals. However the relationships between concentrations in air and soil and those in plants are not well established. Subsequent biotransfer or bioconcentration in domestic animals and humans is also poorly quantified. Many organic chemicals, such as organochlorines, PCB's, dioxins and pesticides are hydrophobic, degrade very slowly in the environment and tend to accumulate in soil and sediment. These chemicals also accumulate in organic phases in plants, such as waxy leaf cuticles or organic matter in roots or stems. Because this route is important for these and other chemicals, an improved predictive capability is desirable. This should be based on a fundamental understanding of the processes involved.

The focus of this paper is on our current work on the air/soil/plant relationships. A preliminary quantitative fugacity-based analysis is presented in the form of both i) an equilibrium, and ii) a kinetic model of uptake and distribution of organic compounds in plants.

Chemicals may enter plants by various routes including :

i)   root uptake followed by possible translocation within the plant;
ii)   foliage uptake from surrounding air vapor;
iii)   contamination and possible penetration of leaf cuticle by chemical sorbed to soil or particulate matter.

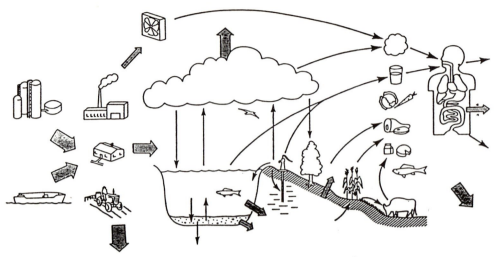

Figure 1.   Some major exposure routes to humans

Plants lose chemicals by:-

i)   Volatilization from leaf surfaces;
ii)   metabolism;
iii)   possibly by diffusion from the roots.

Only the first two pathways of uptake and loss are considered here.

Uptake, accumulation and transport of organic chemicals by plants is largely controlled by the physical chemical properties of a substance such as water solubility, vapor pressure and octanol/water partition coefficients. Correlations have been developed by Briggs et al,[6,7] Bacci et al[8] and Schonherr and coworkers[9] relating concentrations in roots, stem and foliage to concentrations in air and soil using these properties to define the chemical-specific nature of the concentration relationships. Here we explore some of these relationships, using the fugacity format.

FUGACITY

The fugacity concept has been used extensively to model environmental fate of organic chemicals. Recently, it has been applied to the development of a novel plant model[10] to describe the equilibrium distribution of organic chemicals in plants. We present here a modified version of the model which describes the dynamic as well as equilibrium distribution. Briefly, fugacity f (pa) is an equilibrium criterion, related to concentration C (mol/m^3) through a fugacity capacity Z (mol/m^3.Pa) where C = fZ. When two adjacent phases are at equilibrium their fugacities are equal and partitioning can be described in terms of their Z values, that is

$$C_1/C_2 = fZ_1/fZ_2 = Z_1/Z_2$$

The fugacity capacity, or Z value of a chemical is a function of the physical chemical properties of the substance and the characteristics of the phase into which it partitions. A unique Z value is calculated for each chemical in each medium. A summary of Z values is given in Table 1.

In environmental models, D values are assigned as parameters to characterise transport and reaction rates.[2] Chemicals move, between environmental media by diffusive and non-diffusive processes. The diffusive flux between two phases, N_{12} (mol/h), is described by

$$N_{12} = D_{12} (f_1 - f_2)$$

where the parameter D_{12} (mol/Pa.h) is usually a function of a mass transfer coefficient K (m/h), an interfacial area A (m^2) and a Z value (mol/m^3.Pa). Alternatively a diffusivity and path length may be used. The difference between f_1 and f_2 determines the direction of diffusive flow. Movement is from the phase of high fugacity to that of low fugacity. Non-diffusive transfer from phase 1 to 2 is described as

$$N_{12} = G_1 C_1 = G_1 Z_1 f_1 = D_{12} f \text{ mol/h}$$

where G_1 is the volumetric flow rate (m^3/h) of the transporting medium (e.g. sap) and D_{12} has units of mol/Pa.h.

Diffusive and non-diffusive fluxes between phases are summed for a total flux.

First order reaction R (mol/h) in a phase is described as

$$R = k_R V_1 C_1 = k_R V_1 Z_1 f_1 = D_R f_1$$

k_R is a first order rate constant (h^{-1}), V is the phase volume (m^3) and D_R again has units of mol/Pa.h.

Expressing different transport and reaction rates by parameters of identical units assists in their comparison, facilitates grouping and simpliflies the algebra.

Table 1. Definition of Z values (Pa.m^3/mol)

Compartment

Air (Z_A)	1/RT	R = 8.314 (Pa.m^3/mol K)
		T = absolute temperature K
Soil Air (Z_{EA})	1/RT	
Soil Water (Z_W)	1/H or C^S/P^S	H = Henry's Law constant (Pa.m^3/mol)
		C^S = aqueoussolubility(mol/m^3)
		P^S = vapor pressure (Pa)
Soil Organics (Z_{ES})	x.$K_{OC}.\rho_{ES}$/H	K_{OC} = Organic carbon partition coefficient
		= 0.41 K_{OW}
		x = fraction organic carbon = 0.0015
		ρ_{ES} = soil organics density (kg/L)

Soil Mineral (Z_{EM})	not defined	
Plant roots (Z_R)	RCF.$Z_W.\rho_R/\rho_W$	RCF = 0.82 + 0.014K_{OW}
Plant stem (Z_S)	SXCF.$Z_X.\rho_S/\rho_X$	SXCF = 0.82 + 0.0065 K_{OW}
Plant xylem contents (Z_X)	1/H	
Plant phloem contents (Z_P)	1/H	
Plant inner leaf(Z_I)	1/H	
Plant cuticle (Z_{PS})(Z_C)	K_{OW}/H	
Plant fruit, seeds (Z_F)	not defined	
Plant bulk leaf (Z_{BL})	$\emptyset_I Z_I + \emptyset_C Z_C$	

PLANT MODEL

A Qualitative Description

We are are currently working on a project supported by the Natural Sciences and Engineering Research Council of Canada and the Ontario Ministry of the Environment which has the general aim of obtaining experimental data, developing models and devising suitable expressions to describe plant uptake and accumulation from air and soil.

As a result of preliminary efforts to formulate and modify plant models, we believe that the plant, illustrated in Figure 2, should be treated as a minimum of four compartments, roots, stem, inner leaf and leaf cuticle in an environment of bulk air and soil as illustrated in Figure 3. Each compartment or sub-compartment is assumed to consist of volume fractions of air, water and organic matter. The cuticle, in accordance with Kerler and Schonherr[9] is assumed to have properties similar to octanol and therefore to consist of 100% organic matter. Pathways for transport through the plant are through the xylem and the phloem which comprise the plant's vascular system. The xylem conducts the transpiration stream from the root to the foliage, while the phloem conducts nutrients from the site of synthesis in the leaves to growing parts of the plant.

Root uptake

Chemicals in solution in soil water may enter the "apparent free space" in the cortex of the root. This space occupies a volume of 10 - 20% of the root.[11] The major transport process from soil to root for non-ionic xenobiotics is bulk flow. The driving force for xylem sap transport is evapotranspiration of water vapor from the foliage to the atmosphere. Towards the centre of the root is the endodermis. The radial walls of the endodermis are impregnated with suberin, or the Casparian strips, which provide an efficient barrier to water and solute movement. In order to enter the xylem and be transported through the plant, water and chemicals must penetrate the endodermal cells. Chemicals may become sorbed or react there before entering the xylem. Once in the xylem, chemicals are transported through the roots, stem and to the foliage in the transpiration stream. They may react with or partition into various plant parts during transport or they may ultimately be released to the atmosphere in the transpiration stream.

Figure 2.   Illustrative Plant

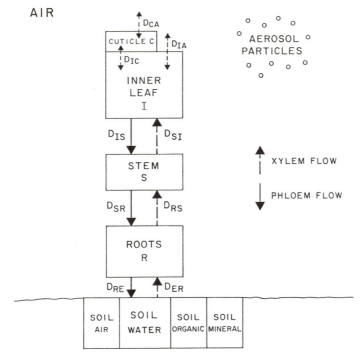

Figure 3.   Schematic illustration of plant model

Foliar uptake

The foliage of the plant is covered by a cuticle which is a waxy or lipid layer made up of a cutin matrix in which are embedded cuticular waxes. The cuticle which is 1 to 10 um thick, provides an effective barrier to transport in and out of the plant. It retards water loss and prevents entry of many pollutants. The cuticle is pierced by stomata, tiny pores about 5 um in diameter, which cover the leaf with a density of approximately 10^8 pores/m^2. The stomata allow transpiration of the evaporated xylem stream and absorption of carbon dioxide required for photosynthesis. They open or close according to environmental conditions.

The inner leaf consists of a substomatal cavity, mesophyll cells which are largely water, and organic matter. Its thickness depends on the plant species but we assume a value of 1 mm here.

Chemicals which enter foliage from surrounding air vapor do so by i) diffusion through the stomata, or ii) absorption to the cuticle where the may become bound or possibly subsequently diffuse through the cuticle to the inner leaf. They may then be transported by the phloem to various parts of the plant.

Exit routes through the foliage include i) diffusion through the stomata in water vapor and ii) diffusion through the cuticle. These processes occur in parallel when the stomata are open.

Quantitative Description

In an attempt to describe the distribution of organic chemicals in plants quantitatively, we have developed a fugacity - based plant model illustrated schematically in Figure 3. Dimensions and volume fractions are assigned to the various compartments and subcompartments as outlined in Table 2. Bulk densities are calculated in terms of their respective volume fractions of air, water and organic matter. The plant dimensions generally resemble those of a 60 day old soybean plant, similar to that which is being investigated experimentally. To describe the partitioning characteristics of the various plant parts, it is necessary to develop correlations for partition coefficients or Z values. This is the subject of separate experimental investigations. The correlations which we use currently are described below.

Fugacity-Capacities

Roots and Stem. The fugacity capacities for air and soil are calculated as described in Table 1. The development of Z values for roots, stem and foliage using a modified form of correlations developed by Briggs et al[6,7] has been described previously;[10] however, a brief summary follows.

Briggs and coworkers[6] developed a root concentration factor, RCF

$$RCF = C_R/C_w$$

where C_R and C_w are chemical concentration in roots (fresh weight) and in external solution respectively, and

$$\log (RCF - 0.82) = 0.77 \log K_{OW} - 1.52$$

Rearranging gives

$$C_R = 0.82C_W + 0.03 (C_O/C_W)^{0.77}.C_W.$$

Where subscript o refers to octanol and K_{OW} is C_O/C_W.

The first term on the right hand side of the equation represents the capacity of the water contained in the root for chemical and the second term represents the capacity of the root tissue. For more hydrophobic chemicals which are of interest here, the second term dominates. For these chemicals, it is possible that equilibrium was not achieved in the 48 hours time period of Briggs' experiment. This would result in under-estimation of RCF. Hypothesising that, on thermodynamic partitioning grounds a power of 1 on K_{ow} applies, results in the correlation

$$RCF = 0.82 + 0.014 K_{OW}$$

Table 2.   Dimensions of Plant compartments and subcompartments

	Volume (m^3)	Thickness or Diameter (m)	Area (m^2)
roots	1.0×10^{-4}	-	-
stem	3.9×10^{-6}	0.005	-
xylem	3.9×10^{-8}	5.0×10^{-4}	-
phloem	2.8×10^{-8}	4.2×10^{-4}	-
inner leaf	1.998×10^{-4}	0.001	0.01
cuticle	2.0×10^{-7}	1.0×10^{-6}	0.01
bulk leaf	2.0×10^{-4}	0.001	0.01

Stem, xylem and phloem are all assumed to be 0.2 m in length.

130

Applying similar non-equilibrium partitioning assumptions to Briggs et al[7] correlation for a stem concentration factor

$$\log(K \text{ stem/xylem sap} - 0.82) = 0.95 \log K_{OW} - 2.05$$

results in a stem/xylem sap concentration factor of SXCF where

$$SXCF = 0.82 + 0.0065 \, K_{OW}$$

RCF and SXCF can be considered to be the equilibrium ratios of chemical concentrations in roots and stems to water and xylem sap respectively and

$$RCF = Z_R \rho_W / (Z_W \rho_R)$$
and $$SXCF = Z_S \cdot \rho_X / (Z_X \cdot \rho_S)$$

where ρ_W, ρ_X, ρ_R, and ρ_S refer to densities of water, xylem sap, root and stem. This correction is necessary since Briggs' correlations consider a mass fraction (g/g) basis, whereas Z value ratios describe a mass/volume ratio (e.g. g/m^3) or mol/m^3.

Therefore the fugacity capacities roots and stem can be expressed as

$$Z_R = RCF \cdot Z_w \cdot \rho_R / \rho_W = (0.82 Z_W + 0.014 Z_o) \, \rho_R / \rho_W$$
and $$Z_S = SXCF \cdot Z_X \cdot \rho_S / \rho_X = (0.82 Z_W + 0.0065 Z_o) \, \rho_S / \rho_X$$

the contents of the xylem or the xylem sap and the phloem, are assumed to have the density and fugacity capacity of water. It seems likely that Z_X and Z_P exceed Z_W because of the presence of organic matter, but the magnitude of the excess is not known and the following relationships are assumed.

$$Z_X = Z_W = 1/H \qquad \text{xylem contents}$$
$$Z_P = Z_W = 1/H \qquad \text{phloem contents}$$

Foliage. The inner leaf is considered to consist of 20% air, 80% water and 2% organic matter and therefore has a fugacity capacity Z_I where

$$Z_I = 0.2/RT + 0.8/H + 0.02 K_{OW}/H$$

The cuticle is assumed to have partitioning properties similar to octanol[9] and a fugacity capacity Z_c where

$$Z_C = Z_O = K_{OW} \cdot Z_W = K_{OW}/H$$

By combining the Z values for inner leaf and cuticle with their appropriate volume fractions $\emptyset_I$ and $\emptyset_c$ we can calculate Z_{BL} for the bulk leaf as

$$Z_{BL} = \emptyset_I Z_I + \emptyset_c Z_C.$$

These fugacity capacities are included in Table 1.

In summary, relatively simple expressions have been developed for the various plant parts which express the fugacity capacity Z in terms of the readily available parameters of molecular weight, solubility, vapor pressure and octanol/water partition coefficients.

D Values

Transport of chemical between and through various plant compartments can be characterized by D values.

In calculating these parameters, we initially define flow rates, G_i (m^3/h), in the xylem and phloem which constitute the major pathways for transport in the plant system. These flow rates are assumed to be uniform throughout the plant.

Crank et al[12] suggested a typical velocity of the transpiration stream to be 20 cm/h. Combining this velocity with the xylem cross sectional area (in this case represented by a diameter of 0.5 mm) results in xylem flow rate G_x, of 4×10^{-8} m^3/h. Flow rates in the phloem are not well quantified but are known to be slow compared to that of the xylem. For the purposes of this model, it is assumed that the phloem flow rate is five percent that of the xylem flow,[11] thus

$$G_P = G_X/20 \; m^3/h.$$

Therefore, bulk flow in the xylem can be characterized by the parameter

$$D_X = G_X.Z_W \; mol/m^3.Pa$$

and in the phloem by

$$D_P = G_P.Z_W \; mol/m^3.Pa.$$

Soil-Root Transfer. Organic chemicals in solution in soil-water enter the root by bulk flow and diffusion, with the former process being considered dominant. The reverse process, from root to soil, is considered to be by diffusion only.

The D value for bulk flow from soil to root will be that of the xylem or D_X. The magnitude of the diffusive flow from root to soil is unknown, but it is arbitrarily assumed to be 5% of the xylem flow.

The D parameters for transport from soil to root are described as

$$
\begin{aligned}
D_{ER} &= D_X & &\text{- soil to root} \\
D_{RE} &= D_{ER}/20 & &\text{- root to soil}
\end{aligned}
$$

Root-stem Transfer. Similarly, transfer from root to stem is considered to take place totally in the xylem resulting in the parameter

$$D_{RS} = D_X \qquad \text{- root to stem}$$

The reverse process from stem to root occurs in the phloem and is described by the parameter

$$D_{SR} = D_P \qquad \text{- stem to root}$$

Stem-Foliage Transfer. Transfer of chemical between stem and foliage is assumed to take place between the stem and the inner leaf compartments by means of the xylem and phloem. This results in parameters for stem-foliage and foliage-stem transfer of D_{SF} and D_{FS} respectively, where

$$
\begin{aligned}
D_{SF} &= D_X & &\text{- stem to foliage} \\
D_{FS} &= D_P & &\text{- foliage to stem}
\end{aligned}
$$

Foliage-Air Transfer. As illustrated in Figure 3, the foliage consists of two compartments, an inner leaf and a cuticle. Chemicals present in the inner leaf may diffuse to the atmosphere via the stomata in the vapor state or through the cuticle.

Inner Leaf to Air via Stomata. Diffusion from inner leaf through the stomata is considered to take place in series through the stomatal pores and the air boundary layer.

Assuming the air boundary layer to have a thickness of 3 mm and the diffusivity of water vapor in air B_W to be 0.08 m^2/h results in a mass transfer coefficient (MTC) for water vapour in air, K_{WB} of.08/.003 or 27 m/h.

By calculating the diffusivity of chemical in air as B_c

$$B_C = B_W.(M/W)^{-0.5} \text{ m}^2/\text{h},$$

the MTC of chemical through the air boundary larger can be calculated as

$$K_{CB} = K_{WB} (B_C/B_W)^{0.67} \text{ m/h}$$

M and W are the molecular weights (g/mol) of the chemical and water respectively.

The D value for transfer through the air boundary layer becomes

$$D_{BL} = K_{CB} A_{FT} Z_A$$

where A_{FT} is the total area of foliage (m^2).

The length of the stomatal opening Y_{ST} is equivalent to the cuticle thickness which in this case is 1×10^{-6} m. In calculating diffusion through the stomata, this path length is corrected for the substomatal cavity according to O'Dell.[13] The estimated stomatal path length Y becomes

$$Y = Y_{ST} + 1/8 \; \pi \; (ab)^{0.5} \text{ m}$$

Where a and b are the major and minor axes of the stomatal opening with values of 1×10^{-5} m and 5×10^{-6} m respectively.

The MTC for water vapor through the stomata becomes

$$K_{WS} = (B_W/Y_{EST}).A_{ST}$$

Where A_{ST} is the stomatal area (m^2/m^2 leaf)

The MTC of chemical through the stomata is calculated as

$$K_{CS} = K_{WS} (B_C/B_W)^{0.67}$$

and the D value for chemical transport through the stomata becomes

$$D_{ST} = K_{CS} \cdot A_{FT} \cdot Z_A$$

The total D value for transport between inner leaf and air, calculated in series becomes

$$D_{IA} = 1/(1/D_{BL} + 1/D_{ST})$$

Inner Leaf to Air via the Cuticle. In calculating diffusion through the cuticle, a diffusivity B_{CC} of 10^{-14} m^2/h[14] is assumed and combined with the diffusion path length Y_C of 0.5 um to obtain a transfer parameter from inner leaf to cuticle, D_{IC}, where

$$D_{IC} = B_{CC}(A_{FT}/Y_C)Z_C$$

The total D value, D_{CA}, for transfer from inner leaf to air then becomes.

$$D_{CA} = 1/(1/D_{IC} + 1/D_{BL})$$

Having defined the volumes, areas, Z values and D values as described previously, it is possible to assemble the overall plant model as shown in Figure 3. It should be noted that the compartments are treated as well mixed volumes. This is, of course, erroneous but it is believed to be an appropriate mathematical simplification in this case. The basic model can be applied and solved in various forms as discussed below.

We are also investigating the most appropriate method of including the deposition of aerosol-associated chemical on to the cuticle surface. An attractive option is to assemble a complete air-soil-plant model and probe the importance of air-soil exchange processes as a means of transferring chemical from soil-to-foliage or air-to-root.

Steady State Version

If constant soil and air fugacities are defined, it is possible to solve the transport and transformation equations to calculate the prevailing fugacities in the various plant compartments, hence the concentrations and amounts of chemical accumulated. Generally the foliage will tend towards a fugacity similar to that of the air and the roots to fugacity equivalent to that of the soil. However, depending on the water solubility and vapor pressure of the chemical, the stem may be influenced variously by the air and soil fugacities. Highly water soluble substances such as 2,4-D tend to be rapidly transported in the sap thus the soil fugacity tends to dominate the fugacity throughout the entire plant. More volatile chemicals tend to be accumulated through the foliage. Analyses by Bacci and coworkers[15] have shown that for certain hydrophobic, but volatile chemicals, transport from the soil to the foliage is easier by soil to air evaporation followed by foliar absorption than it is by root absorption and transport in the sap. This is because the very low solubility of these chemicals in water reduces the Z value in water and hence the D value for sap transport.

It is also possible to calculate the approximate time to achieve equilibrium by comparing the total amount of chemical taken up by a plant compartment at equilibrium and the transport rates. For many chemicals, especially hydrophobic chemicals, this time can be very long, for example years. The steady state analysis is then largely irrelevant in practical terms because of plant growth and death. The steady state analysis is, however, useful in elucidating which compartments are likely to be the primary sites of chemical accumulation and whether uptake from the air or soil is likely to be the dominant route.

Unsteady State Version

The mass balance equations describing the chemical dynamics in the plant can be written in differential equation form and solved numerically to obtain an estimate of the fugacities and concentrations and amounts in the plant at various times Z approaches steady state. If desired, the equations can incorporate varying plant compartment volumes reflecting growth and changing air and soil fugacities. The solutions obtained are of course, specific to the exposure conditions used, thus the numerical solution is very much less convenient than the analytical solution.

In principle, an analytical solution could be obtained using Laplace transforms but it proves to be algebraically very cumbersome.

Approximate Analytical Unsteady State Version

We believe that it should be possible to generate approximate analytical solutions to the differential equations describing the chemical dynamics in the plant. This is the focus of much of our current work in which various schemes are being tried and the approximate solutions compared with the "correct" numerical solutions.

Inspection of the D values applying between various plant compartments enables us to calculate an approximate "lag time" between each compartment. The slow and fast transport processes can be identified and appropriate approximate analytical solutions developed. Often even an approximate lag time is sufficient to demonstrate that a process is unimportant. For example, if the root to stem lag time is one year as may occur for very hydrophobic chemicals, then it is clear that a negligible fraction of chemical will reach the stem during a normal growing season. A very short lag time of perhaps one day between two connected media indicates that they will rapidly reach essentially equal fugacities; thus a steady state approximation may be introduced and a differential equation eliminated.

ILLUSTRATION

To illustrate these concepts, we have gathered data for hexachlorobenzene (HCB) and 2,4-D, extreme examples of hydrophobic and water soluble chemicals and have tested them using the models. Table 3 documents the physical chemical properties of the compounds. The plants were considered to be exposed to chemical in soil and air.

The resulting steady state fugacities and concentrations are compiled in Table 4 as well as the fugacities attributable to air and soil for each compartment. It is interesting to note that the root and stem fugacities are dominated by that of the soil while the foliage fugacity is controlled by that of the air.

Figures 4 and 5, plots of fugacity versus time for the two chemicals in various compartments, illustrate the solution to the differential equations.

As illustrated in Figure 4, HCB with a low water solubility is transported slowly through the plant. Equilibrium between root and stem is not achieved in 6 months, or the period of one growing season in temperate climates. The inner leaf reaches equilibrium with the air in less than one hour due to rapid diffusion through the stomata.

In the case of more water soluble 2,4-D, as shown in Figure 5, the root and stem reach equilibrium in less than 6 months.

Other results such as these can be compared with experimental data and the validity of the various model assumptions tested and the model modified and improved.

Table 3.   Physical Chemical Properties and Z Values for HCB and 2,4-D

	HCB	2,4-D
Physical Chemical Properties		
mol.weight (g.mol)	284.8	221.04
Solubility	0.005	400
Vapour pressure (P_a)	0.0015	1.0
log K_{OW}	5.47	2.81
Z Values		
root	48.4	17.8
stem	22.5	9.08
xylem contents	0.012	1.81
phloem contents	0.012	1.81
inner leaf	0.0094	1.45
cuticle	3450	1168
bulk leaf	3.46	2.62
Air		
gaseous	4.04×10^{-4}	4.04×10^{-4}
aerosols	1.61×10^{6}	4403
soil water	0.012	1.81
soil solids	5.32	1.80

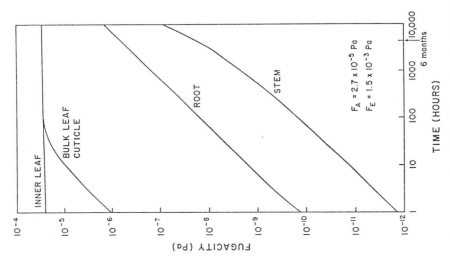

Figure 5.   Plot of fugacity vs time for 2,4-D

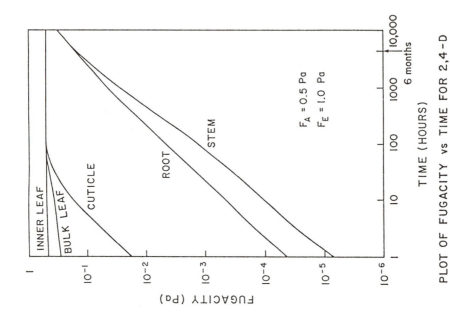

Figure 4.   Plot of fugacity vs time for HCB

136

Table 4. Fugacities and concentrations of hexachlorobenzene (HCB) and 2,4-D in various compartments at steady state

	Compartment Fugacity (Pa)	Fugacity (Pa) Attributable to:-		Concentration (ug/g)
		Air	Soil	
air				
HCB	2.7×10^{-5}	-	-	0.003
2,4-D	0.5	-	-	9.3×10^{4}
soil				
HCB	1.5×10^{-3}	-	-	0.7
2,4-D	1.0	-	-	200
root				
HCB	1.51×10^{-3}	10^{-9}	1.51×10^{-3}	25
2,4-D	1.01	5×10^{-5}	1.01	4770
stem				
HCB	1.5×10^{-3}	10^{-7}		12
2,4-D	1.01	5×10^{-3}	1.0	2440
inner leaf				
HCB	2.7×10^{-5}	2.7×10^{-5}	0	9×10^{-5}
2,4-D	0.5	0.5	10^{-4}	200
cuticle				
HCB	2.7×10^{-5}	2.7×10^{-5}	0	27
2,4-D	0.5	0.5	5×10^{-5}	1.3×10^{5}
bulk leaf				
HCB	2.7×10^{-5}	2.7×10^{-5}	0	0.03
2,4-D	0.5	0.5	10^{-4}	361

CONCLUSIONS

In this paper we have reviewed the current status of our work to develop plant-air-soil models which will be appropriate for calculating the concentrations of chemicals in various plant parts as a result of exposure to contaminated air and soil. We believe that the four compartment plant is a reasonable first approximation which should be capable of yielding results which are suitable for incorporation into human exposure models. Obviously, much work needs to be done to better define the partitioning characteristics, or Z values, and the transport parameters, or D values. Optimal simplifying assumptions must be identified which will enable a model to be developed which is sufficiently simple for general use, yet sufficiently detailed that it will yield useful results and be capable of validation using experimental data.

ACKNOWLEDGEMENTS

The authors are grateful to NSERC for a strategic Grant and the Hazardous Contaminants Coordination Branch, Ontario Ministry of the Environment for funding for this work.

REFERENCES

1.   D. Mackay and S. Paterson, Fugacity Revisited, <u>Environ. Sci. Technol.</u>, 16:654-660 (1982).
2.   D. Mackay, S. Paterson, B. Cheung and E.B. Neely, Evaluating the environmental behaviour of chemicals with a Level III fugacity model, <u>Chemosphere</u>, 14:335-374 (1985).
3.   P.A. Ryan and Y. Cohen, Multimedia transport of particle-bound organics: Benzo(a)pyrene test case, <u>Chemosphere</u>, 15:21-47 (1986).
4.   T.E. McKone and D.W. Layton, Screening the potential risks of toxic substances using a multimedia compartment model: estimation of human exposure, <u>Regul. Toxicol. Pharmacol.</u>, 6:359-380 (1986).
5.   C.C. Travis and A.D. Arms, Bioconcentration of organics in beef, milk and vegetation, <u>Environ. Sci. Technol.</u>, 22:271-274 (1988).
6.   G.G. Briggs, R.H. Bromilow and A.A. Evans, Relationships between lipophilicity and root uptake and translocation of non-ionised chemicals in barley, <u>Pest. Sci.</u>, 13:495-504 (1982).
7.   G.G. Briggs, R.H. Bromilow, A.A., Evans and M. Williams, Relationship between lipophilicity and the distribution of non-ionised chemicals in barley shoots following uptake by roots, <u>Pest. Sci.</u>, 14:492-500 (1983).
8.   E. Bacci and C. Gaggi, Chlorinated hydrocarbon vapours and plant foliage: kinetics and applications, <u>Chemosphere</u>, 16:2515-2522 (1987).
9.   F. Kerler and J. Schonherr, Accumulation of lipophilic chemicals in plant cuticles: prediction from octanol/water partition coefficients, <u>Arch. Environ. Contam. Toxicol.</u>, 17:1-6 (1988).
10.   S. Paterson and D. Mackay, Modelling the uptake and distribution of organic chemicals in plants, <u>in</u>: "Intermedia Pollutant Transport: Modeling and Field Measurements," D.T. Allen and I.R. Kaplan, eds., Plemum Pub. Co. (in press 1989).
11.   P.H. Nye and P.B. Tinker, Solute Movement in the Soil-Root System, Studies in Ecology, Vol. 14, Chapter 5, Blackwell Scientific Pub. (1977).
12.   J. Crank, N.R. McFarlane, J.C. Newby, G.D. Paterson and J.B. Pedley, "Diffusion Processes in Environmental Systems," Macmillan Press Ltd., London, (1981).
13.   R.A. O'Dell, M. Taheri, and R.L. Kabel, A model for uptake of pollutants by vegetation, <u>J. Air Poll. Control Assoc.</u>, 27:1102-1109 (1977).
14.   F. Kerler and J. Schonherr, Permeation of lipophilic chemicals across plant cuticles: prediction from partition coefficients and molar volumes, <u>Arch. Environ. Contam. Toxicol.</u>, 17:7-12 (1988).
15.   E. Bacci and C. Gaggi, Chlorinated pesticides and plant foliage, <u>Bull. Environ. Contam. Toxicol.</u>, 37:850-57 (1986).

AIR-TO-LEAF TRANSFER OF ORGANIC VAPORS TO PLANTS

Eros Bacci* and Davide Calamari[+]

*Department of Environmental Biology
University of Siena
Via delle Cerchia, 3 - 53100 Siena (Italy)
[+]Institute of Agricultural Entomology
University of Milan
Via Celoria, 2 - 20133 Milano (Italy)

INTRODUCTION

In recent years new strategies for the study of the environmental distribution and fate of contaminants were introduced and the research interests have progressively been displaced from the old retrospective and descriptive approaches to the predictive ones. The focus is the understanding of the more significant processes of transformation of chemical substances under "environmental" conditions, as well as the substance partition among the main environmental compartments (e.g.: water, air, soils and sediments, aquatic and terrestrial biomass). The predictive capability of present methods needs to be improved, particularly when complex phenomena have to be predicted in detail: these require complex models, and the knowledge of a great number of parameters.

However, some remarkable improvements have been obtained in the field of "evaluative models" (Baughman and Lassiter, 1978) where the main aim is the comprehension of the key processes and trends in the environmental behaviour of contaminants. It is essential to point out that, in principle, there is no contradiction between the approach of the "complex models", and that of the evaluative models, just as there is no conflict between, for instance, Zoology or Biochemistry and

Ecotoxicology: different strategies are applied for different purposes.

One of the essential tools for the prediction of the environmental behaviour of chemical substances is the possibility of calculation of some environmental equilibrium-partition coefficients from the intrinsic physico-chemical properties, such as water solubility and vapor pressure.

Among the environmental partition coefficients, the bioconcentration factors (BCF), indicating the potential of organisms (or specific tissues) to increase their concentration of a given chemical in respect to that of the surrounding media, are of great ecotoxicological interest. As far as the aquatic environments are concerned, great efforts have been made since the appearance of the first correlation between water solubility and bioaccumulation early in the seventies (Metcalf et al., 1973): in 1974 Neely and coworkers found a significant correlation between BCF in trout muscle and the n-octanol/water partition coefficient (K_{OW}) for 8 different chemicals, with Log K_{OW} ranging from 2.6 and 7.6 (Neely et al., 1974). Since that time, a great number of studies have dealt with refining and improving the BCF/K_{OW} correlation for fish (Veith et al., 1979; Kenaga, 1980).

In terrestrial ecosystems, the research effort to measure and predict analogous BCFs in animals has been slight since for animals the bioaccumulation (from food) is the main entry of lipophylic contaminants into the organisms. In general the air levels of vapors of these contaminants are not able to significantly influence the inputs from food. But also in terrestrial systems there is something taking up chemical vapors and gases from the air to a significant extent, as do the fish from the water: the "green fish", composed by mosses, lichens and higher plant foliage.

PLANT UPTAKE OF NON-POLAR COMPOUNDS

The pathways of a chemical for entering plants are:
- from a contaminated soil via root uptake and translocation;

- from the air, via vapor uptake.

In addition to these main ways, the possibility of direct contamination by wet and dry deposition, followed by retention in external cuticular layers and penetration in plant tissues, and the possibility, for some plants (such as carrots and cress; Topp et al., 1986) of uptake and transport in oil cells have to be taken into account. However, from a general point of view root uptake from the soil and vapor uptake from the air are the main pathways.

The potential of plant foliar tissues to take up non-polar (Log K_{ow} 3-7) and low-volatile (vapor pressure 10^{-1} - 10^{-5} Pa, 20°C, as subcooled liquid) chemicals such as PCB's, DDT or other DDT-like chemicals (e.g.: 2,3,7,8-TCDD, the "Seveso" dioxin) has probably been underrated in some previous works (Suzuki et al., 1977; Bush et al., 1986), and in a recent review (Hansen, 1987). On the other hand, the significance of the air-to-leaf transfer of organics to plants has been pointed out since the sixties (Whitacre and Ware, 1967), and demonstrated by several experimental studies (Nash and Beall, 1970; Bacci and Gaggi, 1985; Bacci and Gaggi, 1986), field data (Iwata and Gunther, 1976; Buckley, 1982; Gaggi et al., 1985), and theoretical approaches.

As reported by Ryan et al. (1988), chemicals with a Log K_{ow} greater than 3.5 are translocated from contaminated soils to stems to a very low extent in soils very poor in organic matter (i.e.: 0.25%), while in soils rich in organic matter (i.e. 6%), are practically no water-mobile. Plant uptake from contaminated soils via root involves a series of consecutive partitions: soil/water, water/root, root/transpiration stream, transpiration stream/stem or foliage. In particular, the transport in the transpiration stream is analogous to a column chromatography (McCrady et al., 1987), where the role of K_{ow} is clearly evident: only chemicals with low Log K_{ow} values (1, 2) can be translocated within the plant (if not degraded), while those with high Log K_{ow} (4 - 7) are strongly sorbed by soil or stopped at the root peel level (Iwata and Gunther, 1976; Weber and Mrozek, 1979; Bacci and Gaggi, 1986).

AN EXPERIMENTAL APPROACH TO MEASURE THE LEAF/AIR BIOCONCENTRATION FACTOR (BCF)

During previous translocation experiments, carried out in small glass green-houses (200-L volume), where plants were grown both in uncontaminated and in contaminated soils, it was observed that the concentration of the vapors of some organochlorine compounds, originating by volatilization from fortified soils, were, within a factor of 2, constant for 4 weeks (Bacci and Gaggi, 1986). This property of the system, mainly due to the constant temperature, to the relatively constant volatilization rate of contaminants, and to a constant air turnover in the green-houses, was then applied to the measurement of the leaf/air equilibrium partition coefficient (BCF) of 10 different chemicals (Bacci et al., 1990).

In Figure 1 a schematic representation of the experimental apparatus is given: azalea (_Azalea indica_, var. Knut Erwèn) plants in pot, with uncontaminated soil are placed in the middle of the glass box, in which vessels containing sand fortified with the chemicals under study are also placed. In a few hours the system reaches a steady state, characterized by relatively constant vapor levels of the different chemicals. A study of the uptake kinetics in plant leaves, using for the analysis only the old leaves, whose growth during the experiment can be assumed as negligible, can be carried out. At the end of the accumulation phase (2-3 weeks), after removing the contaminated soils, the elimination of each chemical taken up by plant leaves can be studied.

An example of experimental results obtained for the herbicide trifluralin is given in Figure 2, where C_A is the concentration of trifluralin vapors in the air (ng g^{-1}; air density = 1.19 g l^{-1}), k_1 and k_2 are, respectively, the first-order uptake- and release-rate constants (h^{-1}) and BCF is the leaf/air dimensionless bioconcentration factor, defined as the ratio between the concentration in the leaf (ng g^{-1}, dry weight) and the level of trifluralin in the air (ng g^{-1}), at equilibrium.

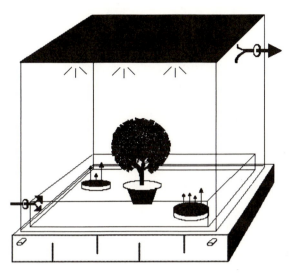

Figure 1. Schematic representation of the apparatus used in the study of uptake and release kinetics of organic vapors by azalea leaves.

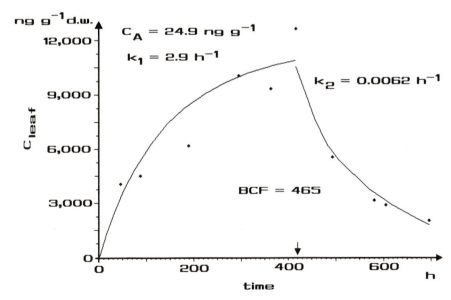

Figure 2. Uptake and release kinetics of trifluralin in azalea leaves.

THE USE OF THE HENRY'S LAW CONSTANT AND K_{ow} IN PREDICTING BCF

Table 1 lists the BCF, Log BCF, H, Log(H BCF), and Log K_{ow} for the chemicals tested with the simple apparatus described above. The few prior studies in this field have usually had different aims, such as the investigation of the accumulation of organic chemicals in growing plants (Topp et al., 1986). A recent research on the partition coefficient between aerial parts of growing sprouts of barley and cress, and air, for five chlorinated benzenes, HCB included, has shown that there is a positive linear Log/Log correlation between BCF and K_{ow} (Scheunert, 1989). Similar results have been reported by Travis and Hattemer-Frey (1988) for p,p'-DDT and DDE, lindane and alpha-HCH, and a PCB mixture (Fenclor 64, analogous to the commoner Aroclor 1260).

If the same correlation is applied to the ten chemicals in Table 1, a poor linear correlation coefficient is obtained (r = 0.86). Better results are obtained if the BCF is assumed to be related to K_{ow} and H, as first suggested by Travis and Hattemer-Frey (1988): the BCF is by definition the ratio of the equilibrium concentration of a chemical in the leaf to its concentration in the air. Dividing the two terms by the equilibrium concentration in water, the BCF becomes:

$$BCF = K_{lw}/K_{aw} \qquad (1)$$

where K_{lw} is the leaf/water partition coefficient and K_{aw} the air/water partition coefficient. The K_{lw} can be obtained from K_{ow}, and K_{aw} is the dimensionless expression of H.

Thus BCF should be directly proportional to K_{ow}/H. This ratio is not easy to use due to the wide range of K_{ow} and H, and to the scarce connection between these two partition properties. So the use of the following equation was proposed (Travis and Hattemer-Frey, 1988):

$$BCF \ H = b \ K_{ow} \qquad (2)$$

where b is a proportionality constant.

Table 1. Azalea leaf/air equilibrium bioconcentration factors, BCF as $(ng\ g^{-1})/(ng\ g^{-1})$, H $(Pa\ m^3\ mol^{-1})$, Log (BCF H), and Log K_{OW} for ten organic chemical vapors (from Bacci et al., 1990).

Compound (or mixture)	BCF	Log BCF	H[*]	Log(BCF H)	Log K_{OW}[**]
thionazin	120	2.08	0.087	1.02	1.2
sulfotep	100	2.00	0.29	1.46	3.0
trifluralin	465	2.67	4.02	3.27	3.0
α-HCH	4,600	3.66	0.87	3.60	3.8
γ-HCH	3,400	3.53	0.13	2.64	3.8
p,p'-DDE	134,950	5.13	7.95	6.03	5.7
HCB	1,860	3.27	131.5	5.39	6.0
p,p'-DDT	192,000	5.28	6.02	6.06	6.0
PCBs (60% Cl)	85,400	4.93	7.11	5.78	6.1
mirex	52,540	4.72	839	7.64	6.9

[*] From Suntio et al. (1988), with the following exception: for the last two chemical H was calculated from vapor pressure and water solubility values taken from the Pesticide Manual (Worthing and Walker, 1983); the PCB mixture was the Fenclor 64, similar to the Aroclor 1260, H from Mackay and Leinonen (1975).

[**] From Suntio et al. (1988), except thionazin and sulfotep, where the values of K_{OW} were calculated by the fragment constant method (Hansch and Leo, 1979).

In this way each chemical can be easily recognised as polar or non-polar. The data reported in Table 1 was processed according to this approach, obtaining the following relationship (Figure 3):

$$Log\ (BCF\ H) = -0.92 + 1.14\ Log\ K_{OW} \qquad r = 0.96 \qquad (3)$$

The slope near to one in the Log/Log relationship indicates that the experimental findings are not in disagreement with Equation (2).

From this relation, chemicals with high lipoaffinity need low H values to be accumulated to a considerable extent, and polar organic compounds may have relatively high BCF values if H is low.

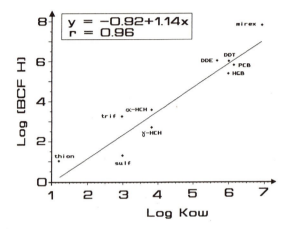

Figure 3. Relationship between the bioconcentration factor of organic chemicals in azalea leaves (BCF), the Henry's Law constant (H) and the n-octanol/water partition coefficient (K_{ow}). From Bacci et al. (1990).

The apparent contradiction inherent the possibility of good correlations between BCF and K_{ow} (without H), could be overcome by the observation that the range of H values of the chemicals studied in these first approaches was relatively narrow (Scheunert, 1989; Asher et al., 1985; Travis and Hattemer-Frey, 1988).

The number of chemicals available to date with a view to predicting the BCF is still very small. For this reason more research is needed before generalizing the results reported in

Figure 3. However, at least for the ten studied compounds, the equilibrium leaf/air bioconcentration factor seems to be controlled by two key partition properties: H and K_{ow}. The former, when high, tends to reduce the bioconcentration potential of higly lipophilic compounds such as mirex or hexachlorobenzene; when small, it may increase the bioaffinity of poorly liposoluble compounds, such as thionazin.

An inverse proportionality between H and the BCF indicates a temperature dependance of the BCF. Increasing the temperature, the vapor pressure of a given chemical generally increases faster than the water solubility, leading to higher H values and consequently lower BCF. This temperature dependance of the BCF could compensate the reduced availability of contaminants in the vapor phase caused by low temperatures: lower vapor pressures correspond to lower volatilization rates and to lower H values, and then to higher BCF. The influence of temperature on BCF certainly is a key point requiring further investigations to improve present knowledge on the quantitative aspects of the partition of organic vapors from plant foliage and air.

REFERENCES

Asher, S.C., Lloyd, K.M., Mackay, D., Paterson, S. and Roberts, J.R., 1985, A critical examination of environmental modelling, in: "Modelling the environmental fate of chlorobenzenes using the persistence and fugacity models", NRC Canada, Ottawa, Canada.

Bacci, E., Calamari, D., Gaggi and C., Vighi, M., 1990, Bioconcentration of organic chemical vapors in plant leaves: experimental measurements and correlation, Environ. Sci. Technol., 24: 885.

Bacci, E. and Gaggi, C., 1985, Polychlorinated biphenyls in plant foliage: translocation or volatilization from contaminated soils? Bull. Environ. Contam. Toxicol., 35:673.

Bacci, E. and Gaggi, C., 1986, Chlorinated pesticides and plant foliage: translocation experiments, Bull. Environ. Contam. Toxicol., 37:850.

Bacci, E. and Gaggi, C., 1987, Chlorinated hydrocarbons vapours and plant foliage: kinetics and applications, Chemosphere, 16:2515.

Baughman, G.L. and Lassiter, R.R., 1978, Prediction on environmental pollutant concentration, in: J. Cairns, Jr., K.L. Dickson and A.W. Maki, eds., "Estimating the hazard of chemical substances to aquatic life", ASTM STP 657, pp. 35-54, American Society for Testing and Materials, Philadelphia (USA).

Buckley, E.H., 1982, Accumulation of airborne polychlorinated biphenyls in foliage, Science, 216:520.

Bush, B., Shane, L. A., Wilson, L.R, Barnard, E. L., and Barnes, D., 1986, Uptake of polychlorobiphenyl congeners by purple loosestrife (Lythrum salicaria) on the banks of the Hudson river, Arch. Environ. Contam. Toxicol., 15:285.

Gaggi, C., Bacci, E., Calamari, D. and Fanelli, R., 1985, Chlorinated hydrocarbons in plant foliage: an indication of the tropospheric contamination level, Chemosphere 14:1673.

Hansch, C. and Leo, A.J., 1979, "Substituent constants for correlation analysis in chemistry and biology". John Wiley, New York (USA).

Hansen, L.G., 1987, Environmental toxicology of polychlorinated biphenyls, in: S. Safe and O. Hutzinger, eds., "Environmental Toxin Series 1", pp. 15-48, Springer-Verlag, Berlin (FRG).

Iwata, Y. and Gunther, F.A., 1976, Translocation of the polychlorinated biphenyl Aroclor 1254 from soil into carrots under field conditions, Arch. Environ. Contam. Toxicol., 4:44.

Kenaga, E.E., 1980, Correlation of bioconcentration factors of chemicals in aquatic and terrestrial organisms with their physical and chemical properties, Environ. Sci. Technol. 14:553.

Mackay, D. and Leinonen. P.J., 1975, Rate of evaporation of low-solubility contaminants from water bodies to atmosphere, Environ. Sci. Technol., 9:1178.

McCrady ,J., McFarlane, C. and Lindstrom, F.T., 1987, The transport and affinity of substituted benzenes in soybean stems, J. Exp. Bot., 38:1875.

Metcalf, R.L., Kapoor, I.P., Lu, P.Y., Schuth, P.Y. and Sherman, P.,1973, Model ecosystem studies of the environmental fate of six organochlorine pesticides, Environ. Health Perspect., 4:35.

Nash, R.G. and Beall, M.L., Jr., 1970, Chlorinated hydrocarbon insecticides: root uptake versus vapor contamination of soybean foliage, Science, 168:1109.

Neely, W.B., Branson, D.R. and Blau, G.E., 1974, The use of the partition coefficient to measure bioconcentration potential of organic chemicals in fish, Environ. Sci. Technol., 8:1113.

Ryan, J.A., Bell, R.M., Davidson, J.M. and O'Connor, G.A., 1988, Plant uptake of non-ionic organic chemicals from soils, Chemosphere, 17:2299.

Scheunert, I., 1989, Correlation of the uptake of organic chemicals by plants from soil with physico-chemical substance properties. Paper presented at the 1st Meeting of the Working Group on Chemical Exposure Prediction of the European Science Foundation (ESF), Milan, 25-26 May, 1989.

Suntio, L.R., Shiu, W.Y., Mackay, D., Seiber, J.N. and Glotfelty, D., 1988, Critical review of Henry's Law constants for pesticides, Rev. Environ. Contam. Toxicol., 103:1.

Suzuky, M., Aizawa, N., Okano, G., Takahashi, T., 1977, Translocation of polychlorobiphenyls in soil to plants: a study by a method of culture of soybean sprouts, Arch. Environ. Contam. Toxicol., 5:343.

Topp, E., Scheunert, I., Attar, A. and Korte, F., 1986, Factors affecting the uptake of ^{14}C-labeled organic chemicals by plants from soils, Ecotoxicology and Environmental Safety, 11:219.

Travis, C.C. and Hattemer-Frey, H.A., 1988, Uptake of organics by aerial plant parts: a call for research, Chemosphere, 17:277.

Veith, G.D., DeFoe, D.L. and Bergstedt, B.V., 1979, Measuring and estimating the bioconcentration factor of chemicals in fish, J. Fish. Res. Board Can., 36:1040.

Weber, J.B. and Mrozek, E., Jr., 1979, Polychlorinated biphenyls: phytotoxicity, absorption and translocation by plants, and inactivation by activated carbon, Bull. Environ. Contam. Toxicol., 23:412.

Whitacre, D.M., and Ware, G.W., 1967, Retention of vaporized lindane by plants and animals, J. Agr. Food Chem., 15:492.

Worthing, C.R. and Walker, S.B., 1983, " The pesticide manual" (7th edition). The British Crop Protection Council, The Lavenham Press Limited, Lavenham, Suffolk (U.K.).

UPTAKE OF ORGANIC CONTAMINANTS BY PLANTS

Craig Mc Farlane

EPA - Environmental Research Laboratory
Corvallis, OR

INTRODUCTION

Plants are central to all nourishment systems in both agro- and natural ecosystems. By fixing energy in carbon molecules (photosynthesis), they serve as the sole source of food and the primary source of minerals for most other organisms. Plants have evolved to efficiently collect CO_2 from the dilute source in the air by providing extensive external and internal surfaces. Water and minerals are efficiently collected from the dilute sources in the soil by intricate root systems that extend to various depths and spread laterally. Roots enhance their own existence by exuding various acids and chelating organic chemicals into their surroundings. This increases mineral solubility and, thus, availability. Roots also provide a rich environment for the growth of numerous microorganisms. Some of these extend the collecting area of the root by becoming saprophytic companions with the roots, and other microbes have enzyme systems that enhance mineralization of decaying matter or accumulate and fix nitrogen from the air. Roots also collect and transport H_2O which carries minerals to the top of plants and provides cooling as well as being a reactant and solvent for almost all biochemical processes.

The fact that plants are so efficient in concentrating energy, water, carbon, and minerals makes them prime candidates for exposure to toxic contaminants. Terrestrial plants may be chemically contaminated by the application of pesticides, accidentally by pesticide drift, or by industrial emissions or spills. Once a chemical is on or in a plant, several different processes may occur. A chemical on the leaves may be photo-degraded, partitioned into the leaf cuticle, mechanically fixed on the surface by leaf hairs or glue-like exudates, or may pass through the cuticle or enter the stomata and move into the cells or be translocated to other parts of the plant. A chemical in the root environment may be adsorbed to the surface, partition between the cell lipids and the bathing polar solution, or be absorbed and translocated in the xylem to other plant organs. Some chemicals are translocated and deposited, others are mobile in plants, and some are excluded. Once in a cell, the chemical is exposed to an array of enzymes that may catalyze reactions that alter the chemical form; degrading some to innocuous compounds and activating others to more toxic forms.

When a chemical is deposited on or is taken up by a plant, there are two concerns. 1) Is the chemical toxic to the plant? And, 2) does it accumulate or persist in a manner which makes it toxic to a consumer? If phytotoxic, the plant dies or is stunted in some aspect of its growth. In an agro-ecosystem, this results in decreased crop yield and related serious economic consequences. In a natural ecosystem, it may change community structure or cause other disruptions in ecosystem processes. If the chemical is non-phytotoxic, it may accumulate in plants or be degraded by them. Both consequences are important in determining contamination of human- and non-managed food chains.

It should be clear from the foregoing that the fate of a chemical in the plant environment is far from simple, and that it depends on both the properties of the chemical as well as the nature of the plant or plant community. Within this array, there are some unifying concepts that have been developed to bring order to our consideration and understanding of these complex relationships. It is my goal to discuss these concepts and present a status report on our progress toward understanding plant contamination. This will include discussion of a model that describes the Uptake, Translocation, Accumulation, and Biodegradation (UTAB) of toxic organic chemicals in terrestrial plants.

PHYTOTOXICITY

Testing

Our work with phytotoxicity has involved the development of several phytotoxicity testing schemes and the compilation of a database useful in risk assessment. The testing protocols were developed for the Office of Toxic Substances (OTS) and include a seed germination/root elongation test[1], and the Early seedling growth test[2]. A new phytotoxicity test is currently being developed to give information previously unavailable with the seed germination and root elongation tests. The test is similar to the young seedling growth test[3] prescribed by the Organization for Economic Cooperation and Development (OECD) in that it includes chemical effect on photosynthesis and transpiration as well as respiration and other enzymatic reactions of plants. The major advantages of this test are: (1) that it is done in hydroponic solution so that the exposure is well defined; whereas, in soil tests (OECD), although the chemical mixture is prescribed, because of soil binding and microbe degradation, the root exposure is not known; and (2) the test is done with mature plants that are dependent on photosynthesis for energy rather than energy reserves in seeds as with seedling tests.

The test has been successfully completed on two species (soybean and barley) using fourteen industrial chemicals with both known and unkown phytotoxicities. The test includes a 5-day exposure to the chemical and yields a dose-response curve useful in risk assessment. It is, thus, timely, and can be performed at low cost with inexperienced personnel.

Database

The database PHYTOTOX was developed by Dr. John Fletcher of the University of Oklahoma and is available from him in a personal computer (PC) version. The database is designed to examine information, rather than simply titles or keywords. The approximately 4000 articles that are currently in the database have each been read and the information cataloged into a computer retrievable format. This yields about 100,000 records each with information about a plant species and the effect of a chemical to some feature of the plant's growth or physiological

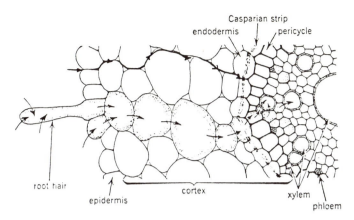

Figure 1. Part of transection of wheat root, illustrating
the kinds of cells that may be transversed by water
and salts and the alternate pathway between cells which
constitutes the apparent free space outward
from the endodermis. Adapted from Esau[21].

functions. Description of the database and examples of its use to
evaluate questions relevant to environmental issues are discussed in a
series of journal articles[4,5,6,7].

UPTAKE OF TOXIC ORGANIC COMPOUNDS BY PLANTS

Root Contamination

The Root Concentration Factor (RCF) defines the partitioning of a
chemical between the external root surfaces (including the apparent free
space) and the exposing solution (typically the soil solution).

$$RCF = \frac{ug/g \ [root]}{ug/g \ [soil \ solution]}$$

Since the outer root cells are not bound by a membrane or other
impervious layer, water and mineral nutrients move passively, using the
sponge-like, amorphous cellulose cell walls as an unhindered passage.
This area has been designated the apparent free space. Anatomically it
is regarded to be the area of the cell walls around and between the
epidermis and cortex cells and bound by the endodermis with its casparian
strips (Fig. 1). The endodermis is important since it provides a barrier
of living membranes through which all water, nutrients and
other solutes must pass. The casparian strip is a layer of suberin
around the endodermis cells that forms an impermeable boundary and
terminates the pathway for water and solutes between cells. This layer
of cells accounts for the semipermeable nature of roots and is the site
where some xenobiotic chemicals are excluded from the plant.

Since the RCF describes the partitioning between a chemical in two
phases (root surfaces and solution), it is natural for it to be similar
to the partitioning of the chemical in other lipid/polar systems. Briggs
et al.[8] showed this relationship to be linear beyond log K_{ow}= 1.2 (Fig.2).

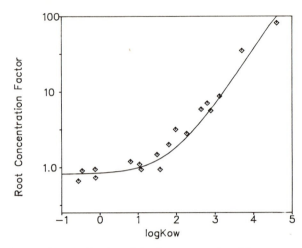

Figure 2. Relationship between the literature
values of the Root Concentration Factors,
and the octan-1-ol/water partition coefficients (as log K_{ow}). Adapted from
Briggs et al.[8].

The RCF is similar in concept to that of the adsorption coefficient,
K_{oc}, for soils[9].

$$K_{oc} = \frac{ug[chemical] \; / \; g[organic\; carbon]}{ug[chemical] \; / \; g[soil\; solution]}$$

The total organic carbon in a soil is commonly expressed as the fraction
of organic carbon. Thus:

$$K_{oc} = \frac{ug[chemical]/g[OC] \; * \; g[OC]/g[soil]}{ug[chemical] \; / \; g[soil\; solution]}$$

The uptake factor (UF) is defined as the ratio of chemical concentration
in roots to the concentration in soil, and by substitution, the
definitions above can be used to predict UF.

$$UF = \frac{RCF}{Koc \; * \; g[OC]/g[soil]}$$

where RCF is determined from the curve presented by Briggs et al.[8] and
the K_{oc} from predictions described by Lyman et al.[9].

The temptation to use an uptake factor (UF) in risk assessment is
understandable and, perhaps, in some cases justifiable. Nevertheless,
there are some things about root contamination that should be considered
before too much credence is given to this value.

First, the RCF applies only to the partitioning of chemicals to surfaces and does not reveal information about the uptake, storage or metabolism of chemicals within the cells of the root. Thus, when considering a fleshy root (i.e., carrot) the outside tissues of the tap root could be expected to follow the predictions of the RCF, but these predictions would not necessarily apply to the inner tissues.

Secondly, the RCF is an equilibrium concept and does not account for events which occur over an extended period. Some studies have determined the RCF for short uptake periods (3-4 h), but it is our experience that at least 24 hours are required for the roots to come into equilibrium with the solution[10]. Subsequent increase of chemical concentration in the roots is caused by storage of the parent chemical, or metabolic products derived therefrom, in cells and tissues remote from the apparent free space and external surfaces of the roots.

Although less obvious, the mass and form of roots associated with individual species are as diversified as the shapes and size of stems and leaves. Some penetrate deeply, while others stay mostly at the surface. Some are characterized by a central tap, and others are a fibrous system. A tap root (i.e., carrot) as seen in the grocery store conceals the existence of an extensive system of fine roots that remains in the soil when harvested. Other soil-borne tissues, such as, potato (a stolon or enlarged stem) or peanut (a fruit) are like roots in that they develop beneath the surface of the soil, but they have dramatically different anatomical features from that of roots. Therefore, their physiological characteristics such as H_2O and mineral uptake are different from that of roots. Thus, in estimating the contamination of the below ground portion of a plant it is essential to consider the anatomy, size, position, and physiology of the organ in question.

UPTAKE AND TRANSLOCATION

In some respects, uptake is the antithesis of chemical binding to roots (RCF). To make the passage through the apparent free space, through the endodermal cells and into the xylem, a chemical must avoid binding and partitioning to the cell surfaces and structural materials. The Transpiration Stream Concentration Factor (TSCF)[11] is a concept useful for describing the process of chemical uptake in the roots and translocation in the xylem to the stem. Values range from 0 (excluded or non-transported chemicals) to 1.0 (uptake and transport at the same rate as water). Values greater than 1.0 are possible if the chemical is actively taken from the soil solution and moved into the transpiration stream faster than water. However, this is a rather rare occurrence, and appears to be restricted to chemicals which are endogenous to plants or synthetic compounds which are similar to naturally occurring substances (i.e. hormones).

$$TSCF = \frac{ug/ml \ [stem \ sap]}{ug/ml \ [soil \ solution]}$$

A value for the TSCF can be determined in several ways. The classical method used to ascertained the TSCF is by determining the rate of chemical movement into the shoot, divided by the rate of water flux (transpiration), as a ratio to the soil solution concentration.

$$TSCF = \frac{(ug/h)/(ml/h)}{ug/ml \ [soil \ solution]}$$

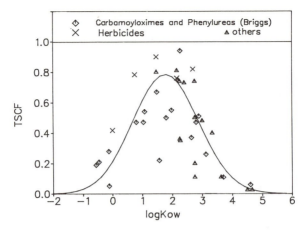

Figure 3. Relationship between the literature
values of Transpiration Stream Concentration
Factor and the octan-1-ol/water partition
coefficients (log K_{ow}). From Briggs et al.[8]
with other unpublished data.

This involves measurement of the chemical in shoot tissues, including an
evaluation of the amount of chemical lost via volatilization and
metabolism, and knowing this in a temporal manner.

A more direct approach for determining the TSCF has been introduced
into and used in our laboratory over the past year. We measure chemical
content in the stem sap. After the plant has come to an equilibrium
condition, sap samples can be obtained from many plants by cutting the
stem with the rooting solution and collecting the stem sap that exudes
due to root pressure or by attaching the stem to a suction device. We
are currently conducting laboratory experiments in which the sap is
collected by attaching a needle to the cut stem and inserting the needle
into an evacuated test tube. However, sap collections are frequently
problematic since water potentials are often too great to allow
collection with suction, and with some stems it is difficult to affect an
air tight connection. Expression of sap under pressure can also be
accomplished with appropriate equipment. However, this is typically
difficult and slow, and the results are confused by the addition of
phloem exudate in the sample.

Still another method of obtaining the TSCF is to determine the loss
of chemical from solution (a much easier task than measuring the flux
into plants). The rate of chemical binding to the roots is subtracted
from this and again divided by the transpiration rate.

$$TSCF = \frac{[(ug/ml\ h)\ *\ ml]\ /\ ml[transpiration]/h}{ug/ml\ [soil\ solution]}$$

This method has the advantage of simplicity since it is not necessary to measure either volatilization or plant metabolism. However a disadvantage lies in the fact that accurate measurement of root and chamber binding is needed. Binding is variable among various chemical treatments and therefore subject to large errors.

Briggs et al.[8] also proposed a structure/activity relationship (SAR) for the TSCF that is represented by a bell shaped curve (Fig. 3) The low TSCF values on the low log K_{ow} end are caused by lower membrane permeability of polar chemicals. On the high log K_{ow} end, the low TSCF values are due to binding of the chemical to surfaces which prevent its translocation within the plant. This concept is generally accepted, although it was quantified on the basis of only a very few observations. In applying this simplified concept, it must be remembered that uptake and translocation involve 1) passage of a chemical by mass flow through the root apparent free space, 2) membrane passage into living cells (symplasm), 3) movement through the endodermis, 4) movement out of the symplasm into the xylem, and 5) mass flow in the transpiration stream through the stems and into the leaves (Fig. 4). En route, the solute is subject to passive movement in the transpiration stream, but also moves within the system by its own energy (diffusion). During passage from soil to plant shoot, a chemical encounters numerous physical surfaces and reaction milieu which may result in chemical partitioning and binding, membrane exclusion, metabolic alteration, and degradation.

In an effort to learn about one major component of this process, we envisioned the xylem, a column of dead cells composed of cellulose, as a chromatographic column. The movement of a series of organic chemicals within this column was examined, and the results treated to a kinetic description[12]. This showed an inverse relationship (Fig. 5) between movement rate and log K_{ow}, thus, supporting the concept proposed by Briggs et al.[8] and allowing us to quantify chemical passage rates.

ACCUMULATION

At the terminal end of the xylem in leaves, water and solutes may move through cell walls, between cells (appoplastic movement) or into the parenchyma cells (symplastic movement). This provides an opportunity for the solute to move into the phloem and be distributed throughout the plant or move by either system to the sub-stomatal cavity and be evaporated. Chemicals with low vapor pressure and no tendency to move into the symplasm (i.e. bromacil) accumulate in the leaves at the ends of the transpiration stream. Others with a high Henry's Law constant are lost to the environment by the same pathway as water[13]. Some chemicals accumulate in and on the root as described by the RCF. Others move into cells and are either stored or metabolically altered. Still others are moved throughout plants, but none are known that are preferentially stored in seeds or other storage organs. Deposition of nutrients and carbohydrates in these organs is via the phloem, and source/sink relationships drive the accumulation. Mechanisms to create such gradients for xenobiotic chemicals are unknown.

BIODEGRADATION

Typically, one thinks of enzymatic degradation of xenobiotic chemicals as being a function of the soil and the bacteria and fungi that reside there. It is also important to understand that plants present a very reactive environment. This was emphasized in a recent study in which we compared the uptake patterns of nitrobenzene in several different species[13].

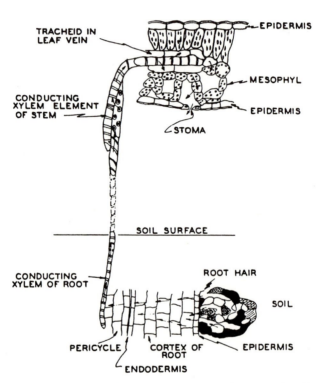

Figure 4. Diagram of the pathway water and solutes take from soil to evaporation from the leaves. From Cox[22].

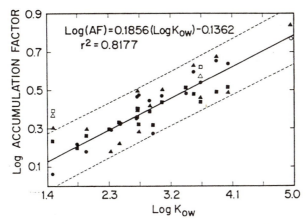

Figure 5. Relationship between the log of the calculated accumulation factor for each chemical and log K_{ow}. The dashed lines represent the 95% prediction intervals. Test 1($\bullet$), test 2($\blacksquare$), test 3($\bigcirc$), test 4($\blacktriangle$), test 5($\square$), test 6($\triangle$).

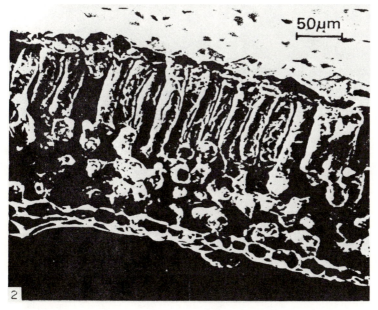

Figure 6a. Scanning electron micrograph of *Phaseolus vulgaris* leaf cross section. From Bole and Parsons[23].

In that study (conducted with ^{14}C labeled nitrobenzene), the TSCF was determined to be about 0.72 for nitrobenzene and was common to all 8 species tested. At the end of the experiment, all plants had similar percent distribution of the tracer in the various compartments of the study (soil solution, plants, and air). The relative distribution of ^{14}C in roots, stems and leaves was similar. At harvest, the tissues were ground and exhaustively extracted with polar (methanol/water) and non-polar (chloroform) extractants. The ^{14}C in each fraction was quantified and, the chloroform fraction was analyzed to verify that it contained the parent nitrobenzene. From **Figure 6** it is clear that only a small fraction (20-30%) of the activity in the roots was still nitrobenzene (non-polar), and that there were significant amounts of metabolic products (polar and insoluble fractions). Nevertheless, it is interesting that all species were similar in regards to root metabolism.

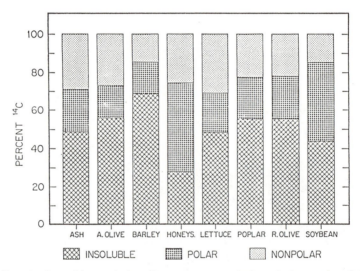

Figure 6b. Carbon-14 activity in roots partitioned into the following fractions: (A) insoluble (B) polar soluble collected in methanol/water, and (C) non-polar soluble collected in chloroform[13].

A similar graph (Fig. 7) for leaf extracts reveals a much different pattern. In the case of honeysuckle, only a small fraction was nitrobenzene, whereas with lettuce and soybeans about 70% remained unaltered. Although no detailed analyses were conducted in this study to identify the products, we presume that the bulk of the H$_2$O soluble ^{14}C was present in glycosylated derivatives of nitrobenzene, since it is well established that this is a common fate of xenobiotics provided to plants[14,15].

These results show that plants have a large capacity to chemically alter nitrobenzene. The capacity to do so is species dependent, and the root and shoot tissue within a single species often differ with regard to their metabolism of nitrobenzene. These facts make it clear that in estimating the fate of xenobiotic compounds in natural habitats, attention should be given to the metabolic properties associated with both root and shoot tissues. Furthermore, chemical analysis of a single species within a plant community will not necessarily indicate how the entire plant community influences the fate of a xenobiotic.

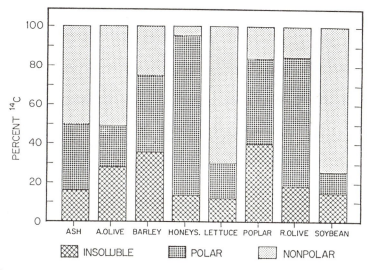

Figure 7. Carbon-14 activity in leaves partitioned into the following fractions: (A) insoluble (B) polar soluble collected in methanol/water, and (C) non-polar soluble collected in chloroform[13].

MODEL

In an effort to support the needs of the EPA's Office of Toxic Substances (OTS) to predict the uptake and fate of chemicals in plants and to supply information for food chain models needed in risk assessment, we are developing a mechanistic, mathematical model[16,17,18,19] to describe uptake, translocation, accumulation, and biodegradation of organic chemicals in plants. The model defines a generic plant as a set of compartments, each representing pertinent plant tissues (Fig. 8). The compartments are separated by boundaries of specified thickness and area, and distinguished by the physical and chemical properties that determine passage of water and solutes. The mass balance equations are solved on a personal computer which also contains example scenarios of plant species and environmental conditions. Our current work is to compare careful laboratory uptake studies with model predictions in an effort to validate the model concepts. These experiments are conducted in special chambers[20] that allow us to follow the movement of a toxic

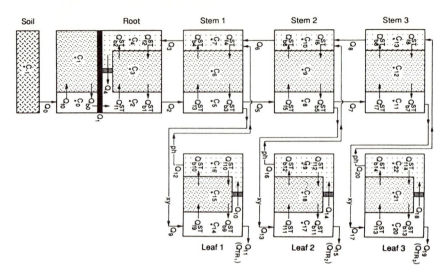

Figure 8. Model diagram showing the adjacent compartments and routes of transfer for water and solutes[24].

chemical while controlling and monitoring the physiology of the plants. The goal is to develop a useful SAR for using the model in a predictive manner. For example, the uptake and distribution of bromacil in plants with three different transpiration rates is compared to experimental data, as shown in Figure 9.

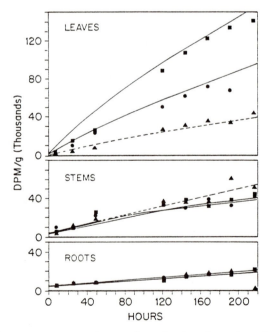

Figure 9. Experimental results (points) and model predictions (lines) obtained at low (▲), medium (●), and high (■) transpiration rates showing concentration of bromacil (^{14}C activity per gram of tissue) in roots, stems, and leaves as a function of time[25].

CONCLUSIONS

Plants are complex organisms which sometimes accommodate and sometimes defy our desire to simplify their interactions with the environment. Our attempt to model these interactions has revealed some generalities, but also some species-specific characteristics. These results have led us to believe that uptake and translocation are similar between species, but that metabolism and phytotoxicity are species-specific.

REFERENCES

1. U.S. Environmental Protection Agency, Seed germination/root elongation toxicity test, EG-12, Office of Toxic Substances, Washington, D.C. (1982).

2. U.S. Environmental Protection Agency, Early seedling growth test, EG-13, Office of Toxic Substances, Washington, D.C. (1982).

3. Organization for Economic Cooperation and Development, Terrestrial plants, Growth test, OECD Guidelines for testing of chemicals, No. 208, Paris, France, (1984).

4. Fletcher, J.S., F.L. Johnson, and C. Mc Farlane, Influence of greenhouse versus field testing and taxonomic differences on plant sensitivity to chemical treatment, Environ. Tox. and Chem. 9:769-776 1990).

5. Fletcher, J.S., F.L. Johnson and J.C. Mc Farlane, Database assessment of phytotoxicity data published on terrestrial vascular plants, Environ. Tox. and Chem. 7:615-622 (1988).

6. Fletcher, J.S., M.J. Muhitch, D. Vann, J.C. Mc Farlane and F. Benenati, PHYTOTOX Database Evaluation of surrogate plant species recommended by the U.S. Environmental Protection Agency and the Organization for Economic Cooperation and Development, Environ. Tox and Chem. 4:523-532 (1985).

7. Royce, L.C., J.S. Fletcher, P.G. Rissler, J.C. Mc Farlane, and F. Benenati, PHYTOTOX: A database dealing with the effect of organic chemicals on terrestrial vascular plants, J. of Chem. Inf. and Computer Sci. 24(7):7-10 (1984).

8. Briggs, G.G., R.H. Bromilow and A.A. Evans, Relationship between lipophilicity and root uptake and translocation of non-ionized chemicals by barley, Pestic. Sci. 13:495-504 (1982).

9. Lyman, W.J., W.F. Reehl, and D.H. Rosenblatt, Adsorption coefficient for soils and sediments, in: "Handbook of chemical property estimation methods (environmental behavior of organic compounds)" McGraw-Hill, NY (1982).

10. Mc Farlane, C., C. Wickliff, Excised barley root uptake of several 14-C labeled organic compounds, Environ. Monito. Assess. 5:385-391 (1985).

11. Shone, M.G.T and A.V. Wood, A comparison of the uptake and translocation of some organic herbicides and a systemic fungicide by barley, J. Exp. Bot. 25:390-400 (1974).

12. McCrady, J.K., C. Mc Farlane and F.T. Lindstrom, The transport and affinity of substituted benzenes in soybean stems, J. Exp. Bot. 38(196):1875-1890 (1987).

13. Mc Farlane, J.C., T. Pfleeger, and J. Fletcher, Effect, uptake and disposition of nitrobenzene in several terrestrial plants, Environ. Tox. and Chem. 9:513-520 (1990).

14. Langebartels, C. and H. Harms, Metabolism of pentachlorophenol in cell suspension cultures of soybean and wheat: pentachlorophenol gulcoside formation, Z. Pflanzenphysiol. 113:201-211 (1984).

15. Ajmand, M. and H. Sandermann, Jr., Metabolism of DDT and related compounds in cell suspension cultures of soybean (*Glycine max* L.) and wheat (*Triticum aestivum* L.), Pest Biochem and Physiol. 23:389-397 (1985).

16. Boersma, L., T. Lindstrom, C. Mc Farlane and E.L. McCoy, Uptake of organic chemicals by plants: A theoretical model. Soil Sci. 146(6):403-417 (1988).

17. Boersma, L., F.T. Lindstrom, C. Mc Farlane, E.L. McCoy, Model of coupled transport of water and solutes in plants, Special Report 818 April 1988, Agricultural Experiment Station, Oregon State University, Corvallis, OR (1988).

18. Lindstrom, F.T., L. Boersma, C. Mc Farlane, K.P. Suen, D. Cawlfield, Uptake and transport of chemicals by plants (Version 2.1), Special Report 819 May 1988, Agricultural Experiment Station, Oregon State University. Corvallis, OR (1988).

19. Lindstrom, F.T., L. Boersma, C. Mc Farlane, Steady state fluid transport model for plants, in: "Irrigation Systems for the 21st Century," Larry G. James and Marshall J. English, eds., American Society of Civil Engineers, New York (1987).

20. Mc Farlane, C., T. Pfleeger, Plant exposure chambers for study of toxic chemical/plant interactions, J. Environ. Qual. 16(4):361-371 (1987).

21. Esau, K., "Plant Anatomy," John Wiley & Sons, Inc. New York (1953).

22. Cox, L.M., The transpiration process as a function of environmental parameters, Thesis, Oregon State University, Corvallis, OR (1966).

23. Bole, G. and E. Parsons, Scanning electron microscopy of the internal cellular structure of plants, J. of Microscopy 98:91-97 (1973).

24. Lindstrom, F.T., L. Boersma, and C. Mc Farlane, Mathematical model of plant uptake and translocation of organic chemicals: Development of the model, J. Environ. Qual. 20:129-136 (1991).

25. Boersma, L., C. Mc Farlane, and F.T. Lindstrom, Mathematical model of plant uptake and translocations of organic chemicals: Application to experiments, J. Environ. Qual. 20:137-146 (1991).

UNCERTAINTIES IN ESTIMATING CHEMICAL

DEGRADATION AND ACCUMULATION IN THE ENVIRONMENT

Stephen T. Washburn and Adam P. Kahn

ENVIRON Corporation
Princeton, N.J.

ABSTRACT

Dwindling landfill capacities, rising landfill costs, and the public health hazards posed by landfill operations have caused many communities to consider incineration of their municipal solid wastes (MSW). Recently, however, concerns regarding the potential human health risks posed by MSW incinerators have fueled widespread opposition to the operation of these facilities. These concerns often focus on the human health risks associated with exposure to contaminants that are emitted from the stacks of MSW incinerators and that may ultimately accumulate in the food chain.

Estimates of the level of exposure to MSW incinerator emissions can be improved by direct measurements of contaminant concentrations in air, soil, plants, livestock, and other media near an operating facility. Such data are usually unavailable, and it is therefore almost always necessary to rely on models to estimate contaminant concentrations in the relevant environmental media. This requires an understanding of the rate and extent of chemical degradation and bioaccumulation in the environment.

In this paper, the uncertainties in estimating chemical degradation and accumulation in the environment are discussed. Examples of the uncertainties in predicting environmental concentrations of chromium and polychlorinated dibenzo-p-dioxins and dibenzofurans are highlighted to illustrate the difficulties that may be encountered in assessing the risks posed by incinerator emissions.

INTRODUCTION

Burning household and commercial garbage, or municipal solid waste (MSW), has been a common practice for centuries. Over the past few decades, however, concerns regarding the potential environmental and human health risks posed by MSW incineration have become increasingly widespread. Initially, these concerns focused on air quality in the immediate vicinity of the incinerators. More recently, investigations have begun to focus on the potential contamination of soil, water, and food.

Municipal Waste Incineration Risk Assessment
Edited by C.C. Travis, Plenum Press, New York, 1991

In order to predict the impact that MSW incinerator emissions can have on air, soil, water, and the food chain, it is necessary to understand the rate and extent of chemical accumulation and degradation in the environment. Unfortunately, characterizing the degradation and accumulation of chemicals in environmental media often involves considerable uncertainty. This uncertainty can arise from:

- The lack of extensive empirical data regarding chemical degradation and accumulation in specific media;

- The lack of a strong theoretical basis for understanding the mechanisms of chemical degradation and accumulation in specific media; or

- The lack of both extensive empirical data and a strong theoretical basis.

Obviously, it would be ideal if both adequate empirical data and a firm theoretical understanding were available. Without empirical data, it is difficult to confirm the validity and applicability of a theoretical model. Without a theoretical basis, it is difficult to understand and evaluate what may appear to be inconsistencies in the empirical data. All too often, scientists are faced with only limited empirical data and an incomplete theoretical basis for predicting chemical degradation and accumulation in the environment.

This paper provides some illustrative examples of the uncertainties that may be encountered in conducting an incinerator risk assessment, especially in predicting the nature of chemical degradation and accumulation. First, we briefly discuss chromium, one of the more hazardous metals emitted from MSW incinerators. We then discuss some of the uncertainties in evaluating food chain exposures to polychlorinated dibenzo-p-dioxins and dibenzofurans, often referred to more simply as dioxins and furans. Dioxins and furans are often the chemicals of greatest concern in incinerator emissions, largely as a result of their relatively high toxicity and their tendency to persist and bioaccumulate in the environment.

CHROMIUM

Chromium is frequently a major contributor to the risk posed by incinerator emissions. Difficulties in estimating the risk associated with food chain exposures to chromium stem from a number of factors, including uncertainties about the oxidation state of chromium in the environment, and limitations in data on the biouptake of chromium by plants.

Like many other metals, chromium can exist in more than one oxidation state in the natural environment, and can interconvert from one state to another. The two forms usually evaluated in an incinerator risk assessment are trivalent and hexavalent chromium. Of these two forms, hexavalent chromium is widely considered to be more toxic. Hexavalent chromium is also believed to have the potential to cause cancer, while trivalent chromium is usually treated as a non-carcinogen. In addition, hexavalent chromium is generally more mobile in the environment than trivalent chromium.

Both trivalent and hexavalent chromium have been detected in MSW incinerator emissions. Obviously, the oxidation state of chromium in the environment is (at least initially) largely dependent upon the form of

chromium emitted from the incinerator stack. As more data from MSW
incinerator emissions tests become available, uncertainties regarding the
oxidation state of chromium in stack emissions decrease. Early incinerator
risk assessments often conservatively assumed that all of the chromium
emitted from the stack would be in the more toxic, more mobile hexavalent
state, and that it would remain in that form in the environment. However,
recent stack test data suggest that less than 10%, and perhaps even less
than 1% of the total chromium emitted from MSW incinerators is in the more
hazardous hexavalent oxidation state. (See, for example, USEPA 1988.)

Despite the fact that we now have more data regarding the form of
chromium emitted from MSW incinerators, uncertainties remain regarding the
oxidation state (and by extension the environmental behavior) of the
chromium once it enters the atmosphere or deposits on soils or plants. For
example, trivalent chromium can oxidize to the hexavalent form under
certain field conditions. Studies suggest that the hexavalent form is
relatively more common in soils that have low organic content and/or
contain electron receptors, such as magnesium oxides; oxygen-rich, moist,
alkaline conditions also tend to favor hexavalent chromium (Bartlett and
James 1979). Thus, site-specific environmental conditions, which are often
unknown or vary in the area impacted by an incinerator, can influence the
oxidation state of chromium in the soil.

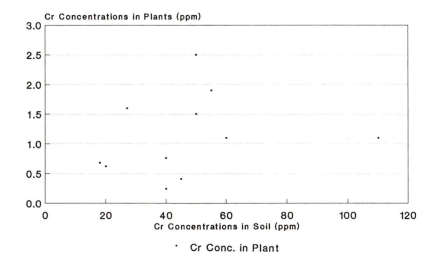

FIGURE 1
UPTAKE OF CHROMIUM BY PLANTS
Source: Baes et al. 1984.

Multi-pathway incinerator risk assessments often consider the human
health risks posed by ingestion of produce grown near the facility. Routes
of possible chromium accumulation in plants include direct deposition onto
exposed surfaces and root uptake from the soil. The contribution from root
uptake is commonly modeled by multiplying the chemical concentration in
soil by a "biouptake factor" to calculate a resulting concentration in the
plant.

Many risk assessments have used biouptake factors developed by Baes et al. (1984) for metals, including chromium (see, for example, ENVIRON 1988, SCAQMD 1988). These biouptake factors were derived primarily from an evaluation of empirical soil-plant uptake measurements; Figure 1 presents the data reported by Baes et al. (1984) for chromium. As shown in the figure, the concentration of chromium in plants varies by about one order of magnitude over the observed concentration range in soil, and the correlation between soil and plant concentration appears to be weak. The relatively wide range in results could be due in part to differences in the oxidation state of the chromium studied in the various tests, or other confounding factors that could not be taken into account.

Fortunately, uncertainties in predicting the oxidation state of chromium in soil, or the extent of plant biouptake of chromium from soil, usually do not have a large impact on the results of an MSW incinerator risk assessment. First, most scientists and regulatory bodies treat both hexavalent and trivalent forms of chromium to be non-carcinogenic via ingestion. Thus, the difference in inherent hazard between the two forms is not nearly as great when considering ingestion pathways as it is when evaluating inhalation exposure. Second, background levels of chromium in soil, typically in the range of 5 to 150 ppm (Schacklette and Boerngen 1984), are often much greater than the concentrations predicted to result from MSW incinerator operations. Chromium that is emitted from an incinerator and deposits onto soil may also tend to reach the distribution between oxidation states that is observed in the background soil. Thus, even though the uncertainty in the root biouptake factor may be substantial, the relative contribution of MSW incinerator emissions to the total amount of chromium taken up by a plant is often small. Third, plant uptake of chromium in soil is only one of several routes by which chromium is accumulated by a plant; other routes (e.g., direct deposition of particulates onto leafy surfaces) may be more significant when evaluating incinerator emissions.

DIOXINS AND FURANS

To a large degree, concerns regarding MSW incinerator emissions have focused on a family of over 200 structurally similar compounds known as polychlorodibenzo-p-dioxins (dioxins) and polychlorodibenzofurans (furans). Chronic exposures to some forms of PCDDs and PCDFs have been linked to birth defects and liver damage (Hatch 1984; Pazderova-Vejlupkova et al. 1981), and has been shown to cause cancer in laboratory animals (IARC 1982). Certain dioxins and furans are also highly persistent in the environment, and have been shown to bioaccumulate in the food chain, especially in fish, meat, and dairy products (Stalling et al. 1983; Jensen et al. 1981; Jensen and Hummel 1982).

When evaluating MSW incinerators sited near bodies of surface water, it is often important to evaluate the potential for dioxin and furan accumulation in fish. The significance of such accumulation can be assessed by first calculating the total input of dioxins and furans to the lake, pond, or river, and then estimating the resulting concentration of dioxins and furans in exposed fish. This second step usually involves the application of a "bioconcentration factor" relating dioxin and furan concentrations in sediment to predicted levels in fish tissue.

Two studies conducted by Kuehl et al. demonstrate the difficulty in developing a bioconcentration factor to describe the biouptake of dioxins and furans from incinerator fly ash. In the first study, Kuehl et al. (1985) exposed carp to two distinct samples of dioxin-contaminated fly ash.

The source of the first sample of fly ash was a large Midwestern MSW incinerator. This ash contained 160 parts per trillion (ppt) of one form of dioxin, 2,3,7,8-tetrachlorodibenzo-p-dioxin (2,3,7,8-TCDD), and had an organic carbon fraction (f_{oc}) of 0.01. The second sample was a composite of fly ash taken from several East Coast MSW incinerators. This ash contained significantly more 2,3,7,8-TCDD (2,000 ppt), in a matrix with an f_{oc} of 0.04. Under identical exposure conditions, carp exposed to the <u>less</u> contaminated Midwest fly ash accumulated greater than 10 times <u>more</u> 2,3,7,8-TCDD than did the carp exposed to the East Coast fly ash (see Table 1).

The results of the experiment are inconsistent with the general expectation that fish exposed to higher levels of 2,3,7,8-TCDD in sediment would exhibit higher 2,3,7,8-TCDD concentrations in their tissue. This can be partially ascribed to the higher f_{oc} of the East Coast fly ash, which would cause the 2,3,7,8-TCDD to sorb more tightly to the ash, and thus be less likely to accumulate in the fish. It is not clear, however, that the four-fold difference in f_{oc} sufficiently explains the very large difference in 2,3,7,8-TCDD concentrations in fish. As noted by the investigators, gill damage suffered by the carp exposed to the East Coast fly ash may also have inhibited dioxin uptake.

TABLE 1
UPTAKE OF 2,3,7,8-TCDD BY CARP
EXPOSED TO INCINERATOR FLY ASH

	FLY ASH		FISH	
Source	2,3,7,8-TCDD Concentrations	f_{oc}	2,3,7,8-TCDD Concentrations	RATIO OF 2,3,7,8-TCDD in Fish to 2,3,7,8-TCDD in Ash
East Coast	2,000 ppt	0.04	2.4 ppt	0.0012
Midwest	160 ppt	0.01	28 ppt	0.175

Source: Kuehl et al. 1985.

A more recent study illustrates another potential source of uncertainty in estimating the uptake of dioxin and furan in fish. In a long-term experiment, Kuehl et al. (1987) exposed carp to 2,3,7,8-TCDD contaminated sediment for a total of 55 days. The level of 2,3,7,8-TCDD in fish tissue was measured periodically over this time. The results of these measurements are plotted in Figure 2.

As shown in Figure 2, the concentration of 2,3,7,8-TCDD in exposed fish continued to increase over the entire period of the experiment. Even after 50 days of exposure, it does not appear that a "steady-state" concentration of 2,3,7,8-TCDD in fish had been reached. While the curve in Figure 2 can be used as a basis for extrapolating to longer periods of exposure, the lack of long-term testing clearly introduces uncertainty into the estimation of "steady-state" 2,3,7,8-TCDD concentrations in fish. It should be noted that a 180-day test of 2,3,7,8-TCDD bioconcentration in aquatic organisms has recently been completed by the USEPA, and that this study may help to reduce some of the uncertainty in predicting long-term concentrations in fish.

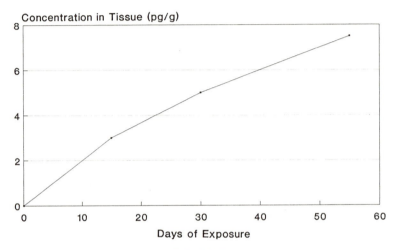

FIGURE 2
UPTAKE OF 2,3,7,8-TCDD BY CARP
EXPOSED TO CONTAMINATED SEDIMENT

The potential contamination of meat and milk is also a serious concern when evaluating the risks posed by some MSW incinerators. In fact, exposure to 2,3,7,8-TCDD through ingestion of cow's milk or beef can represent the greatest potential risk posed by an incinerator (see, for example, Stevens and Gerbec 1988, Connett and Webster 1986). Unfortunately, however, data on the bioaccumulation of dioxins and furans in livestock are limited; this lack of data represents a source of considerable uncertainty in a multi-pathway risk assessment.

The concentration of a chemical in milk is commonly estimated using a "biotransfer factor" similar to the bioconcentration factor used to calculate contaminant concentrations in fish. A biotransfer factor relates the concentration of a contaminant in the milk to the dose of contaminant received by a cow. When possible, biotransfer factors are based on empirical measurements; however, in many instances, they must be estimated based on the physicochemical properties of the contaminant in question.

The biotransfer factor used to predict the extent of 2,3,7,8-TCDD accumulation in milk is often based on tests conducted by Jensen and Hummel (1982). In their experiment, Jensen and Hummel fed three lactating dairy cows fodder containing 2,3,7,8-TCDD concentrations of 500 ppt. Milk was taken from the cows at specific intervals over a 21-day period and analyzed for 2,3,7,8-TCDD. The results of the experiment are summarized in Table 2.

TABLE 2
RESIDUES OF 2,3,7,8-TCDD IN MILK FROM COWS
FED A DIET CONTAINING EXAGGERATED LEVELS
OF 2,3,7,8-TCDD

COW #	TCDD CONCENTRATIONS IN FEED (PPT)	DAYS ON FEED	RESULTING TCDD CONCENTRATIONS IN MILK (PPT)
30	500	21	79
36	500	16	89
7417	500	21	68

Source: Jensen and Hummel 1982.

An examination of the 2,3,7,8-TCDD concentrations ultimately reached in the milk during the tests suggests that the extent of uptake by the three cows tested was roughly similar. The 2,3,7,8-TCDD concentration in milk from cow 7417 at the end of the experiment was 68 ppt; the levels reached in milk from cows 30 and 36 were 79 ppt and 89 ppt, respectively. The 2,3,7,8-TCDD concentrations in milk averaged over the three cows during the experiment are shown in Figure 3; based on such averaging, it would appear that 2,3,7,8-TCDD concentrations in milk were beginning to level off after about 20 days of exposure.

Averaging contaminant concentrations over the three cows can obfuscate potentially important distinctions in uptake behavior by the individual test animals, however. As an illustration, the 2,3,7,8-TCDD concentration data for the three individual cows have been plotted in Figure 4, along with the curve representing the "average." As shown in Figure 4, the milk concentrations for cow 7417 appear to have reached "steady-state" after 21 days. The trend of the curve for cow 36 is less clear, but it is plausible that steady-state concentrations are being approached. The curve for cow 30, however, suggests that 2,3,7,8-TCDD concentrations in milk from this cow are still increasing after 21 days of exposure. Clearly, uncertainties exist in using the data from the three cows to predict the extent of 2,3,7,8-TCDD accumulation in milk.

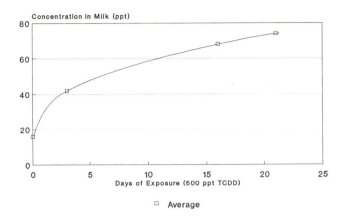

Source: Jensen and Hummel 1982.

FIGURE 3
ACCUMULATION OF 2,3,7,8-TCDD IN MILK

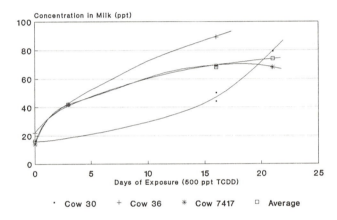

Source: Jensen and Hummel 1982.

FIGURE 4
ACCUMULATION OF 2,3,7,8-TCDD IN MILK

CONCLUSIONS

This paper presents several examples of the difficulties that may be
confronted in conducting an incinerator risk assessment. The uncertainties
inherent in a risk assessment are sometimes substantial, and can limit the
extent to which risks can be accurately quantified. Thus, it is important
to acknowledge the uncertainty of risk assessment, and the inherent
limitations of the data on which it sometimes must be based. It is also
important to recognize that, despite its limitations, risk assessment is
often the best tool for improving our understanding of the potential risks
posed by exposures to toxic chemicals. Risk assessment can also serve to
identify those areas of greatest uncertainty, and to indicate where
additional research should be performed to improve upon our current ability
to quantify risk.

REFERENCES

Baes, G.F., R.D. Sharp, A.L. Sjoreen, and R.W. Shaw. 1984. A review and analysis of parameters for assessing transport of environmentally released radionuclides in agriculture. Oak Ridge National Laboratory, Oak Ridge, TN (ORNL-5786).

Bartlett, R., and B. James. 1979. Behavior of chromium in soils: III. Oxidation. J. Environ. Qual. 8(1).

Connett, P., and T. Webster. 1986. An estimation of the relative human exposure to 2,3,7,8-TCDD emissions via inhalation and ingestion of cow's milk. Presented at the Sixth International Symposium on Dioxin, Fukuoka, Japan. September 16-19.

ENVIRON Corporation. 1988. Site Assessment, Phase 4B: Risk assessment. Prepared for the Ontario Waste Management Corporation. January.

Hatch, M.C. 1984. Reproductive effects of the dioxins. In Public Health Risks of the Dioxins, W.W. Lawrence, ed. Los Altos, California: The Rockefeller University.

International Agency for Research on Cancer (IARC). 1982. IARC Monographs on the evaluation of the carcinogenic risk of chemicals to humans: Chemicals, industrial processses and industries associated with cancer in humans. Vol. 1 to 29, Suppl. 4, 238. Lyon: World Health Organization.

Jensen, D.J., R.A. Hummel, N.H. Mahle, C.W. Kocher, and H.S. Higgins. 1981. A residue study on beef cattle consuming 2,3,7,8-TCDD. J. Agric. Food Chem. 29:265.

Jensen, D.J., and R.A. Hummel. 1982. Secretion of TCDD in milk and cream following the feeding of TCDD to lactating dairy cows. Bull. Environ. Contam. Toxicol. 29:440.

Kuehl, D.W., P.M. Cook, A.R. Batterman, D.B. Lothenbach, B.C. Butterworth, and D.L. Johnson. 1985. Bioavailability of 2,3,7,8-Tetrachlorodibenzo-p-dioxin from municipal incinerator fly ash to freshwater fish. Chemosphere 14(5).

Kuehl, D.W., P.M. Cook, A.R. Batterman, D.B. Lothenbach, and B.C. Butterworth. 1987. Bioavailability of polychlorinated dibenzo-p-dioxins from contaminated Wisconsin river sediment to carp. Chemosphere 16(4):667-679.

Pazderova-Vejlupkova, J., M. Nemcova, J. Pickova, and L. Jirasek. 1981. The development and prognosis of chronic intoxication by TCDD in men. Arch. Environ. Health 36(5).

Schacklette and Boerngen. 1984. Element concentrations in soils and other surficial materials of the conterminous United States, U.S. Geological Survey Professional Paper 1270. Washington, D.C.: United States Government Printing Office.

South Coast Air Quality Management District (SCAQMD). 1988. Multi-pathway health risk assessment, input parameters guidance document. Prepared for the South Coast Air Quality Management District by Clement Associates, Inc. June.

Stalling, D.L., L. Smith. J. Petty, J. Hogan, J. Johnson, C. Rappe, and M. Buser. 1983. Residues of polychlorinated dibenzo-p-dioxins and dibenzofurans in Laurentian Great Lakes fish. In Human and Environmental Risks of Chlorinated Dioxins and Related Compounds. Tucker, R.E., A.L. Young, and A.P. Gray, Eds. Plenum Press. New York.

Stevens, J.B., and E.N. Gerbec. 1988. Dioxin in the agricultural food chain. Risk Analysis 8(3).

United States Environmental Protection Agency (USEPA). 1988. Municipal waste combustion multipollutant study, summary report. Marion County solid waste-to-energy facility. Ogden Martin Systems of Marion, Inc. Brooks, Oregon. EMB Report No. 86-MIN-03A. September.

THE FOOD CHAIN AS A SOURCE OF HUMAN EXPOSURE FROM

MUNICIPAL WASTE COMBUSTION: AN UNCERTAINTY ANALYSIS

Greg D. Belcher[1]
Curtis C. Travis[1]
Randy F. Bruins[2]

[1]Office of Risk Analysis
Health and Safety Research Division
Oak Ridge National Laboratory
Oak Ridge, TN   37831-6109

[2]Environmental Criteria and Assessment Office
U.S. Environmental Protection Agency
26 W. Martin Luther King Drive
Cincinnati, Ohio   45268

INTRODUCTION

The current installed incineration capacity of municipal waste combustors (MWCs) in the United States has reached approximately 50 thousand tons per day and is projected to triple over the next decade.[1] As the use of MWCs as a waste management alternative has increased, public concern over possible environmental and human health effects has also increased. Of particular concern are health risks associated with potential exposures through the food chain.

The food chain is the primary pathway of human exposure for a large class of organics, such as dioxin, PCBs, DDT and other pesticides.[2-4] Because many pollutants emitted by MWCs are lipophilic, extremely persistent compounds, they tend to sorb strongly to air particles, soil, and sediment and to bioaccumulate in living organisms. As a result, the food chain can be a major pathway of exposure to pollutants emitted by MWCs.[5] It is the purpose of this paper to assess the magnitude of human exposure through the food chain for two pollutants released by MWCs: cadmium and 2,3,7,8-tetrachlorodibenzo-p-dioxin (TCDD, commonly referred to as dioxin). These pollutants were chosen as representative of two chemical classes: metals and organics. Cadmium is a metal, which presumably will enter the food chain primarily through vegetative root uptake, while dioxin is an extremely lipophilic compound, which will bioconcentrate in beef and milk.[3]

When models are used as the basis for making estimates of exposure in risk assessment, the question arises as to the sensitivity of model predictions to uncertainties in model input parameters.[6] A first step in the direction of answering this question is the identification of model parameters that make the greatest contribution to overall model error. These represent the parameters that must be determined most accurately in experiments. The second step is to determine probability distributions for each model parameter and propagate these distributions through the model to obtain a characterization of uncertainty in model output. Both of these steps will be attempted for cadmium and dioxin. The extent to which variability in exposure estimates is attributable to uncertainty in the terrestrial food chain model will also be examined.

THE TERRESTRIAL FOOD CHAIN (TFC) MODEL

General Model Description

The TFC Model uses average atmospheric concentration values (ug/m^3) and annual deposition rates ($g/m^2/yr$) of pollutants emitted from MWCs to calculate the human daily pollutant intake through the terrestrial food chain. The model estimates soil concentrations and uptake into vegetation, animals, and humans. Pollutants are assumed to enter the food chain via the following routes:

a. Root Uptake;
b. Direct Deposition;
c. Air-to-Plant Transfer;
d. Direct Ingestion by Man or Animals.

Concentration of Pollutant in Plants

The methodology used to estimate the uptake of chemicals by vegetation and forage crops is given by the following general equation:

$$Cp = (A_d + U_s + T_a) \qquad (1)$$

where Cp = concentration in plants [ug/g Dry Weight (DW)];
A_d = atmospheric deposition component (ug/g DW);
U_s = uptake from soil component (ug/g DW);
T_a = air-to-plant transfer component (ug/g DW).

This equation simply states that the concentration of a pollutant in vegetation results from one of or all of the following three methods: (1) an atmospheric deposition component in which the contaminant is deposited directly onto the plant; (2) a soil uptake component in which the pollutant is first deposited on the soil and then taken up by the roots of the plant and translocated to the edible portion of the vegetation; (3) an air-to-plant transfer component in which the plant absorbs the pollutant from the surrounding contaminated air.

Atmospheric Deposition Component. The atmospheric deposition component is calculated by:

$$A_d = \frac{K2 \bullet D_y \bullet (1 - F_v) \bullet R_p \bullet [1.0 - \exp(-K_p \bullet T_p)]}{Y_p \bullet K_p} \qquad (2)$$

where A_d = atmospheric deposition component (ug/g DW);
$K2$ = units conversion factor of 1000;
D_y = annual deposition rate of pollutant (g/m^2/year);
F_v = fraction of pollutant in the gaseous phase (unitless);
R_p = interception fraction of the edible portion of the plant (unitless);
K_p = plant surface loss coefficient (1/years);
T_p = time period of exposure to deposition per harvest of the edible portion of the plant (years);
Y_p = yield or standing crop biomass of the edible portion of the plant (kg DW/m^2).

Uptake from Soil Component. Vegetative uptake from soil is given by:

$$U_s = B_s \bullet Sc \qquad (3)$$

such that

$$Sc = \frac{K1 \bullet D_y \bullet (1 - F_v) \bullet [1.0 - \exp(-K_s \bullet T_c)]}{Z \bullet Bd \bullet K_s} \qquad (4)$$

where U_s = uptake from soil component (ug/g DW);
B_s = soil-plant bioconcentration factor (ug/g DW)/(ug/g soil);
Sc = soil concentration of pollutant after the total time period of deposition (ug/g soil);
$K1$ = units conversion factor of 100;
D_y = annual deposition rate of pollutant (g/m^2/yr);
F_v = fraction of pollutant in the gaseous phase (unitless);
K_s = soil loss constant (1/years);
T_c = period of long-term buildup in soil (years);
Z = soil depth (cm);
Bd = soil bulk density (g/cm^3).

The soil loss constant, K_s, is the sum of all soil losses due to leaching, K_l, or degradation and volatilization, K_{dv}, and therefore:

$$K_s = K_l + K_{dv}. \qquad (5)$$

The formula for calculating the soil loss constant for losses due to leaching is :

$$k_l = \frac{P + I - Ev}{Theta \bullet 15.0 \bullet [1.0 + (Bd \bullet Kd/Theta)]} \qquad (6)$$

where k_l = loss constant due to leaching (1/years);
     P   = average annual precipitation (cm/year);
     I   = average annual irrigation   (cm/year);
     Ev = average annual evapotranspiration (cm/yr);
     Theta = soil volumetric water content (ml/cm^3);
     15.0 = soil depth from which leaching removal occurs (cm);
     Bd = soil bulk density (g/cm^3);
     Kd = soil water partitioning coefficient (ml/g).

    Air-to-Plant Transfer Component.   The air-to-plant transfer component is given by:

$$T_a = \frac{B_v \bullet (C_v + C_y \bullet F_v)}{Da} \qquad (7)$$

where T_a = air-to-plant transfer component (ug/g DW);
     B_v = air-to-vegetation bioconcentration factor (ug/g DW)/(ug/g air);
     C_v = concentration of pollutant in air due to volatilization from soil $(ug/m^3$ air);
     C_y = average atmospheric concentration of pollutant due to direct emission from a municipal waste incinerator $(ug/m^3$ air);
     F_v = fraction of pollutant in the gaseous phase (unitless);
     Da = density of air $(1190$ g/m^3 at 25 °C).

    Concentration of Pollutant in Animal Tissues. The estimation of the concentration of the pollutants in food and forage crops is just the initial phase in calculating total exposure from the food chain. The concentrations in animal tissues and milk that result from animals ingesting contaminated forage and grains must also be determined. Pollutant concentration for each animal tissue group is modeled via:

$$A_i = (Qf_i \bullet C_f + Qg_i \bullet C_g + Qs_i \bullet Sc) \bullet Ba_i \qquad (8)$$

where A_i = concentration of pollutant in the ith animal tissue group (ug/g DW);

    Qf_i = quantity of forage eaten by the ith animal type each day (kg DW/day);
    C_f = concentration of pollutant in animal forage (ug/g DW);
    Qg_i= quantity of locally grown grain eaten by the ith animal type each day (kg DW/day);

C_g = concentration of pollutant in animal grain feed
(ug/g DW);

Qs_i = quantity of soil eaten by the ith animal type each
day (kg soil/day);

Sc = soil concentration of pollutant after the total
time period of deposition (ug/g soil);

Ba_i = biotransfer factor for the ith animal tissue group
(ug day/kg DW).

Human Daily Intake of Pollutants

Human Daily Intake of Pollutant due to Direct Ingestion
of Contaminated Soil. Three routes exist by which humans can
ingest pollutants: (1) direct ingestion of contaminated soil;
(2) consumption of contaminated plants; and (3) consumption of
contaminated animal tissues. Direct ingestion of pollutants
can occur from intentional eating or inadvertent hand-to-mouth
transfer. The food chain model calculates human daily
pollutant intake resulting from ingestion of contaminated soil
by:

$$Is = Sc \bullet Cs \bullet E \qquad (9)$$

where Is = total human daily pollutant intake due to direct
ingestion of contaminated soil (ug/day);

Sc = soil concentration of pollutant after the total
time period of deposition (ug/g soil):

Cs = soil ingestion rate (g/day);

E = exposure duration adjustment (unitless). Soil
ingestion generally occurs in children from the
ages of 1 to 6 years. The exposure duration
adjustment factor is used to more effectively
spread this exposure over the entire
life span when dealing with carcinogenic
pollutants. This parameter is derived by dividing
the duration of the exposure by the assumed life
span. However, "E" should be set equal to 1.0 when
determining the adult human total daily pollutant
intake.

Human Daily Intake of Pollutant due to Consumption of
Contaminated Plant Tissue. Pollutants can be ingested by
consuming contaminated crops. The total amount of pollutants
ingested each day as a result of consumption of an individual
plant group is calculated as:

$$Ip_i = Cp_i \bullet Fp_i \bullet Ep_i \qquad (10)$$

where Ip_i = human daily intake due to consumption of the ith
plant group (ug/day);

Cp_i = concentration of pollutant in the ith plant group
(ug/g DW);

Fp_i = fraction of ith plant group that is locally
produced (unitless);

Ep_i = daily dietary consumption of the ith plant group
(g DW/day).

Human Daily Intake of Pollutant due to Consumption of Contaminated Animal Tissue. Total amount of human daily pollutant intake due to consumption of a single, contaminated animal tissue is given by:

$$Ia_i = A_i \bullet Fa_i \bullet Ea_i \qquad (11)$$

where Ia_i = total human daily pollutant intake resulting from consumption of the ith animal tissue group (ug/day);

A_i = concentration of pollutant in the ith animal tissue group (ug/g DW);

Fa_i = fraction of the ith animal tissue group that is locally produced;

Ea_i = daily human consumption of the ith animal tissue group (g DW/day).

Total Human Daily Pollutant Intake. Total human daily pollutant intake is calculated by:

$$It = Is + \sum_{i=1}^{np} Ip_i + \sum_{i=1}^{na} Ia_i \qquad (12)$$

where It = total human daily pollutant intake (ug/day);

Is = total human daily pollutant intake due to direct ingestion of contaminated soil by humans (ug/day);

Ip_i = the total human daily pollutant intake resulting from consumption of all contaminated plant tissues (ug/day);

np = number of plant groups;

Ia_i = the total human daily pollutant intake resulting from consumption of all contaminated animal tissues (ug/day);

na = number of animal groups.

PROBABILISTIC APPROACH

Uncertainty Analysis

When models such as the TFC Model are used to estimate levels of human exposure, the question arises as to the accuracy of model predictions. The use of conservative assumptions have predominated in the past, estimating exposure levels based on worst case scenarios.[6] However, as the emphasis has moved to eliminate conservative assumptions and increase the "realism" of model predictions, uncertainty analysis has become an essential requirement in determining the reliability of model predictions.[7]

Figure 1 generally describes the methodology of an uncertainty analysis. As the figure shows, a model exists as a function of a series of input parameters. Each of these parameters can be described as a random variable with an appropriate probability distribution. Random samples are

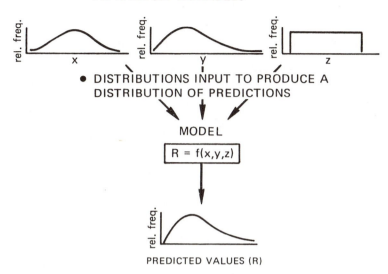

Figure 1.   General Approach to Uncertainty Analysis

Table 1. % Contribution of the Most Influential Parameter
         to Variability in Total Daily Intake

Chemical	Most Influential Parameter	Percent Influence
Cadmium	**Dy**	57.0%
Dioxin	**Dy**	55.8%

taken from these distributions and are input to the model to produce a single estimate of model output. This process is repeated many times to produce a distribution of model predictions. Hence, as opposed to a point estimate that has no characterization of model variability, an uncertainty analysis produces a range of model estimates from which a best estimate (i.e. mean, median, etc.) can be ascertained.

A computer code called PRISM was used to evaluate the uncertainty analysis associated with model predictions resulting from model input parameter variability. PRISM uses a systematic sampling method to generate random model parameter values and iteratively simulates the model predictions for each parameter set. PRISM analyzes the results to determine means, variances, and important statistical relationships between model predictions and input parameters.[8]

The systematic sampling methodology employed by PRISM is Latin Hypercube Sampling (LHS). This technique generates a sample from a given distribution by dividing the total distribution into a number of equal segments and selecting a value at random from within each segment. This constrained random sampling procedure forces sampling from the tails of the distribution, resulting in a small sample being more representative of the distribution than unconstrained Monte Carlo, or simple random sampling, techniques.[9] In fact, when efficiency is defined as the minimum number of samples needed to sufficiently represent the probability distributions of each parameter, LHS appears to be the most efficient systematic sampling methodology. In comparison, LHS requires approximately 10 times fewer samples than simple random sampling techniques.[8] One-hundred fifty repetitions of the LHS process were performed for each chemical considered.

For the uncertainty analysis of the TFC Model, proper probability distributions were determined for each model input parameter. These distributions were propagated through the model to obtain a characterization of model output. The PRISM code allows for the input of several types of distributions. However, the code is designed to take only normal distributions as input and to convert those numbers to the type of distribution specified by the user.

The general approach used to determine the appropriate uncertainty distributions for the model parameters is described in this section. For more information, Section IV provides the details on how each individual parameter distribution was obtained. The general approach used included the following basic methods for making judgments about appropriate uncertainty distributions:

o   Using previously published model parameter uncertainty distributions. Previous research has been performed for some of the parameters used in the TFC model, with input distributions for use in stochastic modeling already created.

o   Using data from primary source literature to develop
    appropriate probability distributions.

o   Plotting parameter values on probit paper to determine
    if the parameter adhered to a lognormal distribution.

o   Assuming a lognormal distribution if data limitations
    made distribution type undiscernible.

The reader is encouraged to review the model parameter
uncertainty distributions used in this analysis, keeping in
mind that distribution selection was a subjective judgment
based on the current state of knowledge. It is also important
to note that, according to R.H. Gardner (co-author of the
PRISM code), while the selection of distribution type is an
important component of the parameter description in the model,
for 90% of the time the mean and variance are the most
critical moments for purposes of stochastic modeling.[10]

Sensitivity Analysis

     One goal of the uncertainty analysis of the TFC Model was
to determine those model parameters that make the greatest
contribution to overall model output variability. The most
effective process to determine parameter influence is to use
an uncertainty analysis based on sensitivity analysis.
Sensitivity analysis is defined as quantifying the magnitude
of a given input parameter's effect on model predictions.[11]
Therefore, a sensitivity analysis was performed using the
capabilities of PRISM, which outputs percentage rankings of
each parameter's influence on each model output as part of the
code's uncertainty analysis. The sensitivity analysis of the
TFC Model was therefore performed as part of the uncertainty
analysis as described in Section III(A) with parameters varied
over their likely ranges. The results of the sensitivity
analysis, as summarized in Table 1, show that variability in
the annual deposition rate ($g/m^2/yr$), Dy, is the primary
contributor to uncertainty in model estimates of total daily
intake for cadmium and dioxin.

TECHNICAL BASIS FOR MODEL PARAMETER VALUES

     The TFC Model parameters used in this uncertainty
analysis are either chemical-independent or chemical-specific.
This section describes the technical basis for each parameter
distribution. The independent and dependent parameters are
discussed separately.

Chemical-Independent Parameters

     The chemical-independent parameter values and
distributions are summarized in Table 2. The technical

Table 2. Chemical-Independent Parameters

Parameter	n	Dist.[a] Type	Normal Mean	Normal St. Dev.	Range	Reference(s)
BD	299	1	1.40	0.15	0.93 to 1.84	12
Ca[b]						
Beef	4	1	33.9	2.2	31.7 to 35.8	13,14
Beef liver	4	1	6.4	0.7	5.4 to 7.0	13,14
Lamb	4	1	1.7	0.2	1.4 to 1.8	13,14
Pork	4	1	10.7	1.4	8.5 to 11.6	13,14
Poultry	4	1	11.9	1.7	10.2 to 14.1	13,14
Dairy	4	1	40.9	5.5	33.0 to 45.3	13,14
Eggs	4	1	7.0	1.0	6.2 to 8.1	13-15
Cp[b]						
Potatoes	4	1	30.0	1.9	27.4 to 32.0	13,14
Leafy Veg.	4	1	2.6	0.3	2.4 to 3.0	13,14
Fresh Leg.	4	1	0.8	0.2	0.6 to 1.1	13,14
Dried Leg.	4	1	3.3	0.8	2.4 to 4.4	13,14
Root Veg.	4	1	3.3	0.3	3.0 to 3.6	13,14
Garden Fr.	4	1	11.0	1.0	9.7 to 12.0	13,14
Grains	4	1	177.9	4.5	171.2 to 180.7	16-19
Cs	4	1	0.037	0.043	0.0044 to 0.10	
E		c	1.0			
Evap	69	1	64.81	17.98	32.50 to 100.00	13
Fa						
Beef	4	1	0.0526	0.0056	0.0462 to 0.0599	20
Beef Liver	4	1	0.0526	0.0056	0.0462 to 0.0599	20
Lamb	4	1	0.0526	0.0056	0.0462 to 0.0599	20

184

Parameter	n	Dist.[a] Type	Normal Mean	Normal St. Dev.	Range	Reference(s)
Fa						
Pork	4	1	0.0526	0.0056	0.0462 to 0.0599	20
Poultry	4	1	0.0719	0.0197	0.0439 to 0.0893	20
Dairy	4	1	0.0403	0.0057	0.0347 to 0.0482	20
Eggs	4	1	0.0644	0.0185	0.0427 to 0.0870	20
Fp						
Potatoes	4	1	0.0931	0.0405	0.0503 to 0.1288	20
Leafy Veg.	4	1	0.0931	0.0405	0.0503 to 0.1288	20
Fresh Leg.	4	1	0.0931	0.0405	0.0503 to 0.1288	20
Dried Leg.	4	1	0.0931	0.0405	0.0503 to 0.1288	20
Root Veg.	4	1	0.0931	0.0405	0.0503 to 0.1288	20
Garden Fr.	4	1	0.0765	0.0286	0.0500 to 0.1171	20
Grains	4	1	0.0044	0.0041	0.0011 to 0.0102	20
Irrigat	69	1	32.99	27.08	12.50 to 100.00	14
Precip	69	1	88.18	34.74	18.06 to 164.19	21
Of						
Beef	7	1	3.60	1.36	2.12 to 5.74	22-25
Beef Liver	7	1	3.60	1.36	2.12 to 5.74	22-25
Lamb	3	1	1.58	0.07	1.50 to 1.64	22-25
Pork		c	0.0			25,26
Poultry		c	0.0			25,26
Dairy	7	c	10.37	2.42	7.28 to 15.18	22,24
Eggs		c	0.0			25,26

(Continued)

185

Table 2. (Continued)   Chemical-Independent Parameters

Parameter	n	Dist.[a] Type	Normal Mean	Normal St. Dev.	Range	Reference(s)
Qg						
Beef	7	1	4.05	1.53	2.39 to 6.46	22-25
Beef Liver	7	1	4.05	1.53	2.39 to 6.46	22-25
Lamb		c	0.0			22
Pork	2	1	2.07	0.050	2.03 to 2.10	24,25
Poultry	5	1	0.053	0.13	0.038 to 0.066	24,25
Dairy	7	1	4.03	0.94	2.83 to 5.90	22-24
Eggs	5	1	0.053	0.13	0.038 to 0.066	24,25
Qs						
Beef	4	1	0.39	0.27	0.1 to 0.72	24
Beef Liver	4	1	0.39	0.27	0.1 to 0.72	24
Lamb	3	1	0.04	0.03	0.016 to 0.07	26
Pork	3	1	0.034	0.034	0.0 to 0.0688	26,27
Poultry	3	1	0.001	0.001	0.0 to 0.002	26,27
Dairy	5	1	0.41	0.24	0.1 to 0.72	24
Eggs	3	1	0.001	0.001	0.0 to 0.002	26,27
Rf	12	n	0.47	0.30	0.02 to 0.82	12
Rg		c	0.0			5
Rp						
Potatoes		c	0.0			
Leafy Veg.	9	1	0.16	0.10	0.08 to 0.38	14,22
Fresh Leg.	2	1	0.008	0.004	0.005 to 0.01	14,22
Dried Leg.	2	1	0.008	0.004	0.005 to 0.01	14,22
Root Veg.		c	0.0			

Parameter	n	Dist.[a] Type	Normal Mean	Normal St. Dev.	Range	Reference(s)
Rp						
Garden Fr.	6	l	0.05	0.05	0.004 to 0.008	14,22
Grains	2	c	0.0			19
Tc	2	l	65.00	49.50	30.00 to 100.00	14
Tf	2	l	0.123	0.058	0.082 to 0.164	14
Tg	2	l	0.164	0.083	0.082 to 0.247	12
Theta	299	n	0.22	0.07	0.03 to 0.40	14
Tp	2	l	0.164	0.083	0.082 to 0.247	12
Yf	10	l	0.31	0.25	0.02 to 0.75	22
Yg	48	l	0.30	0.09	0.14 to 0.45	
Yp						
Potatoes	2	l	0.48	0.106	0.405 to 0.555	14,22
Leafy Veg.	9	l	0.177	0.086	0.091 to 0.353	14,22
Fresh Leg.	2	l	0.104	0.38	0.077 to 0.130	14,22
Dried Leg.	2	l	0.104	0.38	0.077 to 0.130	14,22
Root Veg.	5	l	0.334	0.142	0.090 to 0.434	14,22
Garden Fr.	6	l	0.107	0.093	0.012 to 0.253	14,22
Grains	48	l	0.30	0.09	0.14 to 0.45	14,22
Z	2	l	20.0	5.00	15.0 to 25.0	19

[a] Distribution types: l = lognormal, n = normal, c = constant.
[b] Distribution types for $\underline{Ca}$ and $\underline{Cp}$ based on McKone and Ryan.[24]

derivation and/or justification of each parameter distribution is described below.

BD: Soil Bulk Density. Data were taken from Hoffman and Baes.[12] This document summarizes values for BD, which have been extensively measured and documented in the literature. Hoffman and Baes conclude that soil bulk density appears to be lognormally distributed based on a probability plot. The arithmetic mean of the 299 reported values is 1.40 (g/cm^3).

Ca: Daily Adult Consumption of Meats. Data were taken from wet weight (WW) values reported by Yang and Nelson[13] and converted to dry weight (DW) values using conversion factors from Baes et al.[14] and USDA.[15] Distributions were assigned to daily intakes of beef, beef liver, lamb, pork, poultry, milk, and eggs. The respective (g DW/day) arithmetic mean intakes are as follows: 33.9, 6.4, 1.7, 10.7, 11.9, 40.9, 7.0. Lognormal distributions were assumed.

Cp: Daily Adult Consumption of Plant Food Groups. Yang and Nelson's[13] reported WW values were converted to a DW basis using conversion factors given by Baes et al.[14] Distributions were designated lognormal and were assigned to human intakes of potatoes, leafy vegetables, fresh legumes, dried legumes, root vegetables, garden fruits, and grains. The respective arithmetic mean (g DW/day) intakes are as follows: 30.0, 2.6, 0.8, 3.3, 3.3, 11.0, 177.9.

Cs: Soil Ingestion Rate for Adults. Soil ingestion data for adults is sparse.[16,17] In order to obtain a distribution for the parameter, best estimates of Eschenroeder[16], Lagoy[17], Kimbrough[18], and the U.S. EPA[19] were used as data to calculate an arithmetic mean of 0.037 (g/day) of soil incidentally ingested by adults. Due to the limited data, a lognormal distribution was assumed.

E: Exposure Duration Adjustment Factor. This parameter is used in the TFC Model to effectively spread soil ingestion exposure over the entire life of an individual when dealing with carcinogenic pollutants. However, E is set equal to 1.0 when evaluating toxic effects other than carcinogenesis and when determining the adult human total daily pollutant intake.

Evap: Average Annual Evapotranspiration. Values were extrapolated from Baes et al.[14] for 69 selected cities across the United States. An arithmetic mean of 64.81 (cm/year) was calculated; however, due to data limitations, probability plots were unable to ostensibly discern a distribution type, and a lognormal distribution was assumed.

Fa: Fraction of Meat Group that is Raised Locally. Data were taken from U.S. EPA[19] based on the average percent of seasonal consumption that is homegrown for various food groups, all households included. Distributions were determined from the reported percentages for spring, summer, fall, and winter, and they were assumed to be lognormal.

Distributions were assigned for the following meat groups: beef, beef liver, lamb, pork, poultry, dairy, and eggs. The respective arithmetic means (unitless) were as follows: 0.0526, 0.0526, 0.0526, 0.0526, 0.0719, 0.0403, and 0.0644. The available data grouped beef, beef liver, lamb, and pork together, such that their resulting distributions are the same.

Fp: Fraction of Plant Group that is Raised Locally. Data were taken from U.S. EPA[20] based on the average percent of seasonal consumption that is homegrown for various food groups, all households included. Distributions were determined from the reported percentages for spring, summer, fall, and winter, and they were assumed to be lognormal. Distributions were assigned for the following plant groups: potatoes, leafy vegetables, fresh legumes, dried legumes, root vegetables, garden fruits, and grains. The respective arithmetic means (unitless) were as follows: 0.0931, 0.0931, 0.0931, 0.0931, 0.0931, 0.0765, 0.0044. The available data grouped potatoes, leafy vegetables, legumes and root vegetables together, such that their distributions are the same.

Irigat: Average Annual Irrigation. Values were extrapolated from Baes et al.[14] for 69 selected cities across the United States. Due to data limitations, probability plots were unable to discern a distribution type for this parameter, and it was assumed to have a lognormal distribution with an arithmetic mean of 32.99 (cm/yr).

Precip: Average Annual Precipitation. Values were taken from the U.S. Bureau of the Census[21] for 69 selected cities across the United States. Annual totals were given in inches and converted to centimeters. A probability plot of the parameter showed that the average annual precipitation appears to be lognormally distributed. The arithmetic mean of Precip was 88.18.

Qf: Quantity of Forage Eaten by an Animal. Data were gathered for beef, beef liver, lamb, and dairy.[22-25] The respective arithmetic means (kg/day) were as follows: 3.60, 3.60, 1.58, 10.37. A lognormal distribution was assumed. Pork and poultry were assumed not to ingest forage crops.[25,26]

Qg: Quantity of Locally Grown Grain Eaten by an Animal. Data were gathered for beef, beef liver, pork, poultry, dairy, and eggs.[22-25] The respective arithmetic means (kg/day) were as follows: 4.05, 4.05, 2.07, 0.053, 4.03, 0.053. Lognormal distributions were assumed, and sheep were assumed not to ingest grain.[22]

Qs: Quantity of Soil Eaten by an Animal. Data were gathered for each animal group.[24,26,27] The arithmetic means (kg/day) were as follows: beef (0.39), beef liver (0.39), lamb (0.04), pork (0.034), poultry (0.001), dairy (0.41), eggs (0.001). Lognormal distributions were assumed.

Rf: Interception Fraction of the Edible Portion of Animal Forage Crops. Measured values of RF found in the literature are summarized by Hoffman and Baes.[12] Values listed are averages taken from 12 reports. Hoffman and Baes show that the probability distribution of reported values more nearly approximates a normal instead of lognormal distribution. The arithmetic mean (unitless) of reported values is 0.47.

Rg: Interception Fraction of the Edible Portion of Animal Grain Feed. Since grains have "protective" husks, they are considered similar to protected produce and assumed not to be contaminated via deposition of facility-emitted pollutants.[5,28] Thus, this variable is assumed constant at 0.0 (unitless).

Rp: Interception Fraction of the Edible Portion of the Plant Food Group. Interception fractions were calculated from the relationships for leafy vegetables and exposed produce listed in Baes et al.[14] Potatoes, root vegetables, and grains are assumed equal to zero. Garden fruits and legume values were calculated from the exposed produce equation. Fresh weight standing crop biomass values, which were used as inputs into the equations, were taken from Shor et al.[22] A probability plot of the parameter values for leafy vegetables showed the values fit a lognormal distribution. The arithmetic means (unitless) by plant group are as follows: leafy vegetables (0.16), fresh and dried legumes (0.008), garden fruits (0.05).

Tc: Total Time Period of Deposition. **Tc** equals the lifetime of a combustor. U.S. EPA[19] estimates the lifespan of a municipal waste combustor to range from 30 to 100 years. These values were used for the range of the parameter, which is assigned a lognormal distribution.

Tf: Exposure Time to Deposition Per Harvest of Animal Forage Crops.   Baes et al.[14] reports values for the average time between successive hay harvests (60 days) and the average time between successive grazings by cattle (30 days). These values were divided by the number of days in a year (365) to get a range of values for the parameter with an arithmetic mean of 0.123 (years). A lognormal distribution was assumed.

Tg: Exposure Time to Deposition Per Harvest of the Edible Portion of Grain Used for Animal Feed. Reported values for the average time between successive hay harvests, 60 and 90 days[14], were divided by the number of days in a year (365) to obtain a range of values for the parameter, with an arithmetic mean of 0.164 (years). A lognormal distribution was assumed.

Theta: Soil Volumetric Water Content. Values for the parameter have been extensively reviewed in the theoretical literature and are summarized by Hoffman and Baes[12]. Hoffman and Baes determined the distribution of **theta** to be more normally than lognormally distributed based on probability plots of 299 data points. The arithmetic mean of the distribution is 0.22 (ml/cm^3).

Tp: Exposure Time to Deposition Per Harvest of the Edible Portion of Plants Consumed by Humans. Reported values for the average time between successive harvests, 60 and 90 days[14], were divided by the number of days in a year (365) to obtain a range of values for the parameter, with an arithmetic mean of 0.164 (years). A lognormal distribution was assumed.

Yf: Yield or Standing Crop Biomass of the Animal Forage Crop. Hoffman and Baes[12] summarized the data from 10 studies, determining that a lognormal distribution was appropriate. The arithmetic mean of the reported values was 0.31 (kg DW/m^2).

Yg: Yield or Standing Crop Biomass of the Edible Portion of Animal Grain Feed. Data were taken from Shor et al.[22] for the 48 contiguous states. The reported values were therefore averages of the standing crop biomass for each state. A probability plot of the 48 values showed that Yg fit a lognormal distribution. The arithmetic mean of the parameter is 0.30 (kg DW/m^2).

Yp: Yield or Standing Crop Biomass of the Edible Portion of Plant Groups Consumed by Humans. Data were determined from Baes et al.[14] and Shor et al.[27] Fresh weight data listed in Shor were converted to dry weight using the conversion factors from Baes. A lognormal distribution was determined based on a probability plot for leafy vegetables. Arithmetic means (kg DW/m^2) of the parameter for each plant group are as follows: potatoes (0.48), leafy vegetables (0.177), legumes (0.104), root vegetables (0.334), garden fruits (0.107), grains (0.30).

Z: Soil Depth. Values for soil depth normally range from 15 to 25 centimeters.[19] Hence, a lognormal distribution was assumed using an arithmetic mean of 20.

Chemical-Specific Parameters

Tables 3 and 4 summarize the distributions and give references for each parameter, by chemical.

Ba: Biotransfer Factor for the Animal Tissue Group. The biotransfer factor (BTF) is defined as the equilibrium concentration of pollutant in an organism or tissue (mg/kg) divided by the average daily intake of pollutant (mg/day).[23] However, a traditional measure of a chemical's potential to accumulate in animal tissue is the bioconcentration factor (BCF), which is defined as the equilibrium concentration of pollutant in an organism or tissue (ug/g) divided by the equilibrium concentration of pollutant (ug/g) in food for terrestrial organisms. Hence, when values were reported in terms of BCFs, they were converted to BTFs by dividing the BCF by the dry feed ingestion rate (kg/day) for the given animal

Table 3. Cadmium: Chemical-Specific Parameters

Parameter	n	Dist.* Type	Normal Mean	Normal St. Dev.	Range			Reference(s)
Ba								
Beef	8	1	0.003	0.005	0.0	to	0.012	27
Beef Liver	6	1	0.135	0.129	0.045	to	0.3793	27
Lamb	6	1	0.027	0.026	0.003	to	0.06	27
Pork	6	1	0.028	0.048	0.0	to	0.12	27
Poultry	12	1	12.45	11.94	1.36	to	36.57	27
Dairy	3	1	0.0055	0.002	0.0036	to	0.0076	27
Eggs	3	1	0.6496	0.1836	0.4894	to	0.8500	27
Bf	3	1	0.39	0.17	0.22	to	0.55	14,34,35
Bg	8	1	0.05	0.07	0.003	to	0.22	35
Br								
Potatoes	4	1	0.09	0.03	0.06	to	0.14	27,35
Leafy Veg.	6	1	1.18	0.72	0.04	to	2.38	27,35
Fresh Leg.	4	1	0.24	0.32	0.03	to	0.70	27,35
Dried Leg.	4	1	0.24	0.32	0.03	to	0.70	27,35
Root Veg.	11	1	1.98	4.757	0.0072	to	18.76	27,35
Garden Fr.	5	1	1.16	1.36	0.0019	to	3.14	27,35
Grains	8	1	0.05	0.07	0.003	to	0.22	35
Bv								
Potatoes		c	0.0					
Leafy Veg.		c	0.0					
Dried Leg.		c	0.0					
Fresh Leg.		c	0.0					

Parameter	n	Dist.* Type	Normal Mean	Normal St. Dev.	Range	Reference(s)
Bv						
Root Veg.		c	0.0			
Garden Fr.		c	0.0			
Grains		c	0.0			
Bvf		c	0.0			
Bvg		c	0.0			
Cy	NA	l	4.0E-6	5.31	1.3E-7 to 1.3E-4	16,30,36-44
Dy	NA	l	7.3E-6	7.61	1.1E-7 to 4.7E-4	16,30,36-44
Kd	3	l	148.13	238.08	6.55 to 423.0	14,35
Kp	4	l	31.88	14.83	18.07 to 52.51	14,35
Ksqv		c	0.0			

* Distribution Types: l = lognormal; c = constant.
NA = not applicable.

Table 4. Dioxin: Chemical-Specific Parameters

Parameter	n	Dist.[a] Type	Normal Mean	Normal St. Dev.	Range	Reference(s)
B_a						
Beef	3	1	0.3089	0.4285	0.0550 to 0.80	23,45,46
Beef Liver	3	1	0.8097	1.1256	0.0286 to 2.10	23,45-47
Lamb[b]	1	1	0.0213			28
Pork[b]	1	1	0.5610			28
Poultry[b]	1	1	118.96			28
Dairy	3	1	0.030	0.008	0.024 to 0.039	46
Eggs[b]	1	1	90.03			28
B_f	6	1	0.033	0.034	0.00003 to 0.083	23,24,48
B_g	6	1	0.033	0.034	0.00003 to 0.083	23,24,48
B_r						
Potatoes	6	1	0.033	0.034	0.00003 to 0.083	23,24,48
Leafy Veg.	6	1	0.033	0.034	0.00003 to 0.083	23,24,48
Fresh Leg.	6	1	0.033	0.034	0.00003 to 0.083	23,24,48
Dried Leg.	6	1	0.033	0.034	0.00003 to 0.083	23,24,48
Root Veg.	6	1	0.033	0.034	0.00003 to 0.083	23,24,48
Garden Fr.	6	1	0.033	0.034	0.00003 to 0.083	23,24,48
Grains	6	1	0.033	0.034	0.00003 to 0.083	23,24,48
B_v						
Potatoes		c	0.0			
Leafy Veg.	3	1	1.02E4	7.65E2	9.61E3 to 1.10E4	49-52
Dried Leg.	3	1	1.02E4	7.65E2	9.61E3 to 1.10E4	49-52

194

Parameter	n	Dist.[a] Type	Normal Mean	Normal St. Dev.	Range	Reference(s)
Bv						
Fresh Leg.	3	l	1.02E4	7.65E2	9.61E3 to 1.10E4	49-52
Root Veg.		c	0.0			
Garden Fr.	3	l	1.02E4	7.65E2	9.61E3 to 1.10E4	49-52
Grains		c	0.0			
Bvf	3	l	1.02E4	7.65E2	9.61E3 to 1.10E4	49-52
Bvg		c	0.0			
Cy	NA	l	9.1E-11	13.46	4.4E-13 to 1.9E-8	53
Dy	NA	l	3.5E-11	15.49	1.3E-13 to 9.8E-9	53
Kd	3	l	3.1E4	8.48E3	2.3E4 to 4.0E4	54,55
Kp	2	l	27.06	12.84	17.98 to 36.14	2,27
Ksqv	2	l	0.0624	0.0061	0.0578 to 0.0693	49,54

[a]Distribution Types: l = lognormal, c = constant.
[b]Estimated based on background concentration of TCDD in these tissues.
NA = not applicable.

group.[23] If only one bioconcentration factor was available, a distribution of biotransfer values was created by dividing by the minimum, arithmetic mean, and maximum of the appropriate dry feed ingestion rates. The dry feed ingestion rates used to convert to BTFs were determined by adding the quantity of grain (Qg) and forage (Qf) eaten by the appropriate animal group. Lognormal distributions were assumed.[24]

Bf: Plant-Soil Bioconcentration Factor for the Animal Forage Crop. This parameter was recorded in terms of ($\underline{u}$g pollutant/g plant tissue DW) / ($\underline{u}$g pollutant/g soil). **Bf** is a measure of a chemical's potential to accumulate in plant tissue and is defined as the equilibrium concentration of the pollutant in forage crops divided by the equilibrium concentration of the pollutant in soil. A lognormal distribution was assumed.[24]

Bg: Soil-to-Plant Bioconcentration Factor for Grain Used for Animal Feed. This parameter was recorded in terms of ($\underline{u}$g pollutant/g plant tissue DW) / ($\underline{u}$g pollutant/g soil). **Bg** is defined as the equilibrium concentration of the pollutant in animal grain feed divided by the equilibrium concentration of the pollutant in soil. A lognormal distribution was assumed.[24]

Br: Soil-to-Plant Bioconcentration Factor for the Vegetation Groups Consumed by Humans. **Br** was recorded in terms of ($\underline{u}$g pollutant/g plant tissue DW) / ($\underline{u}$g pollutant/g soil). **Br** corresponds to the **Bf** and **Br** parameters for animal feed, except that it is a measure of a chemical's tendency to accumulate in the tissues of vegetation consumed by humans. A lognormal distribution was assumed.[24]

Bv: Air-to-Plant Bioconcentration Factor for the Vegetation Groups Consumed by Humans. This parameter is defined as the equilibrium concentration of the organic in upper plant parts ($\underline{u}$g pollutant/g plant tissue DW) divided by the concentration of the organic in air as a vapor ($\underline{u}$g pollutant/g air). By definition, metals are assumed not to encounter air-to-leaf transfer.[28] Thus, this parameter was set constant to 0.0 for all cadmium. Since data for air-to-leaf transfer factors are sparse, **Bv**'s for dioxin were estimated from the following geometric mean regression equation for organics developed by Travis and Hattemer-Frey[28]:

$$Bv = 5.0 \bullet 10^{-6} \, K_{ow} \div H \qquad\qquad (13)$$

where K_{ow} = the organic's octanol water partitioning coefficient; and H = the Henry's Law Constant for the organic.

Bv was applied in the model to the following food groups: leafy vegetables, legumes, and garden fruits. A lognormal distribution was assumed based on a probability plot of 9 values of **Bv** calculated for dioxin from the above equation.

Bvf: Air-to-Plant Bioconcentration Factor for the Animal Forage Crop. Bvf corresponds to the Bv parameter discussed previously, except that it applies to animal forage crops instead of human vegetative crops. Again, values for cadmium were assumed constant at 0.0.

Bvg: Air-to-Plant Bioconcentration Factor for the Animal Grain Feed. Bvg corresponds to the Bv and Bvf parameters discussed previously, except that it applies to grains used for animal feed. However, since grains are considered to be protected by protective shells, they were considered to be uncontaminated via this pathway.[28] Therefore, this value was considered constant at 0.0 for all chemicals.

Cy: Atmospheric Concentration. This parameter reflects the average atmospheric concentration of the pollutant due to direct emissions from a municipal waste combustor and is recorded in terms of (ug pollutant/m^3 air). The methodology used to determine Cy is described in Hattemer-Frey and Travis[28] and Lowe et al.[29] First of all, average annual atmospheric concentrations resulting from a unit release of 1 gram of pollutant per second were calculated for 24 operating facilities across the United States (based upon a 50-kilometer radius of each facility).[30] Facility characteristics and locations can be found in Table 5.

Next, a range of emission rates was determined for each chemical (Table 6). These distributions of emissions were merged with the exposure concentration distributions based on a unit emission rate to obtain a distribution of atmospheric concentrations for each chemical. For a more elaborate description of how this procedure has been used for determining distributions of atmospheric concentrations of pollutants emitted from MWCs, the reader is referred to Hattemer-Frey and Travis[28] and Lowe.[29] Also, since atmospheric concentrations were determined for an "arbitrary" MWC based on data from incinerators throughout the United States, estimates for Cy can be expected to vary more than site-specific estimates of average annual atmospheric concentrations. Lognormal distributions were assumed.[28]

Dy: Annual Dry Deposition Rate. The methodology used to determine distributions of the annual dry deposition rate (g/m^2/yr) for each pollutant corresponds to the procedure used to determine distributions for Cy. The Industrial Source Complex Long-Term (ISCLT) deposition model was used to estimate average annual dry deposition rates for each of the 13 chemicals within a 50-km radius of the 24 MWCs used in this analysis. These distributions of deposition rates were merged with the distributions of emissions for each chemical (Table 6). This procedure resulted in a distribution of deposition rates. Also, since annual dry deposition rates were determined for an "arbitrary" MWC based on data from incinerators throughout the United States, estimates for Dy can be expected to vary more than site-specific deposition rates. Lognormal distributions were assumed.[28]

Table 5. Operating Characteristics and Location of the
24 MWCs used to Determine Unit Emission Results
for C_y and D_y*

Location	Stack Height (m)	Stack Diameter (m)	Exit Gas Velocity (m/s)	Exit Gas Temperature (°K)
Livingston, MT	5	1.8	3.5	420
Ogden, UT	46	1.5	13.4	477
Collegeville, MN	37	3.6	0.7	350
Waukesha, WI	32	2.1	9.2	420
Stuttgart, AZ	11	0.8	9.3	420
Palestine, TX	15	0.9	5.5	350
Pinellas Co., FL	49	3.7	9.2	420
Lewisburg, TN	15	1.2	6.6	350
Baltimore, MD	96	1.8	13.4	506
Euclid, OH	38	0.8	7.3	505
Harrisburg, PA	49	0.6	8.3	420
Huntington, NY	38	2.1	12.1	494
Durham, NH	35	1.8	11.5	455
Bellingham, WA	11	1.1	6.2	420
Meredith, NH	24	0.9	11.1	830
Pittsfield, NH	8	0.6	21.2	420
Oswego Co., NY	43	0.6	22.1	420
Johnsonville, SC	9	2.1	1.8	420

Location	Stack Height (m)	Stack Diameter (m)	Exit Gas Velocity (m/s)	Exit Gas Temperature (°K)
Ames, IA	61	2.4	1.3	505
Haverhill, MA	99	2.1	19.0	435
Waxahachie, TX	27	0.9	11.8	350
Lakeland, FL	61	3.7	1.3	400
Sitka, AK	24	1.0	4.0	420
Osceola, AR	9	1.2	5.6	420
GEOMETRIC MEAN	26.5	1.4	6.6	435

*Sources: U.S. EPA.[56]; Hattemer-Frey and Travis.[28]

NOTE: Geometric Means include all MWCs listed in Table 5.

Table 6. Predicted Emission Rates

Chemical	Geometric Mean (g/s)	Geometric Standard Deviation
Cadmium[a]	1.06E-3	0.76
Dioxin[b]	2.39E-8	1.24

[a]Emission rates for cadmium based on estimated values for 11 proposed MWCs in the United States.[16,36-44,57]

[b]Emission rates for dioxin based on actual emissions from existing MWCs in the United States.[28,56]

Table 7. Breakdown of Exposure Pathways for Cadmium
(% contributions to average daily intake)

1. Direct ingestion of contaminated soil :   12.9

2. Consumption of contaminated vegetation:

       Atmospheric deposition component   3.3
       Root Uptake component   62.8
       Air-to-plant transfer component   0.0

3. Consumption of contaminated meat and
          dairy products:

       Soil-Animal-Human Route   13.3
       Plant-Animal-Human Route   7.7

Kd: Soil-to-Water Partitioning Coefficient. **Kd**
represents the ratio of elemental concentration of soil to
that in water in a soil-water system at equilibrium. The
parameter was recorded in terms of (ml/g).   Baes et al.[14]
assigned a lognormal distribution for this variable based on
distributions determined for cesium and strontium.

Kp: Plant Surface Loss Constant. **Kp** is expressed in
units of (1/years). The equation for calculating the **Kp** is:

(ln 2) / (half-time in # of days / 365 (days/year))   (14).

This parameter was assigned a lognormal distribution based on
Miller and Hoffman's results.[31]

Ksgv: Soil Loss Constant. This parameter represents the
soil loss constant due to degradation (biotic and abiotic) and
volatilization (1/years).   Metals were assumed not to incur
soil loss, and this parameter was therefore set constant at
0.0 for cadmium.   The equation for calculating **ksgv** for
organics is:

(ln 2) / (half life in soil in # of days / 365 days)   (15).

A lognormal distribution was assumed.

RESULTS

Cadmium

    Table 7 shows that cadmium intake due to emissions from
MWCs is due mainly to the consumption of contaminated
vegetation (66.1%).   Almost 63% of total daily intake is
attributed to the root uptake component, while atmospheric
deposition contributes only 3.3%.   The analysis also
determined the breakdown of cadmium intake according to
general food groups. Table 8 shows that garden fruits (32.3%)
and root vegetables (19.6%) are the primary plant tissue
groups contributing to total daily intake (TDI).

    The consumption of contaminated meat and dairy products
is also an important pathway of cadmium exposure (21.0%) as
shown in Table 7. Table 9 shows the percentage contributions
to daily intake of each meat food group, with beef liver
offering the greatest influence (12.3%).   This particular
pathway can be further subdivided into plant-animal-human and
soil-animal-human.   The latter path is the most prolific,
contributing more than 13% of total cadmium intake.

    A final goal of the uncertainty analysis was to determine
a probability distribution for the daily human intake of
cadmium due to emissions of the contaminant from MWCs. Figure
2 shows that these values fit a lognormal distribution, and
their statistical summary is given in Table 10.   Figure 3

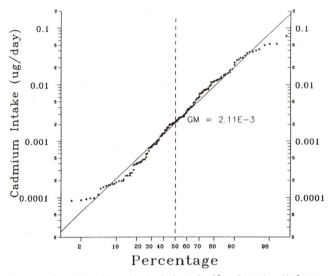

Figure 2.　Predicted Cadmium Daily Intake Values

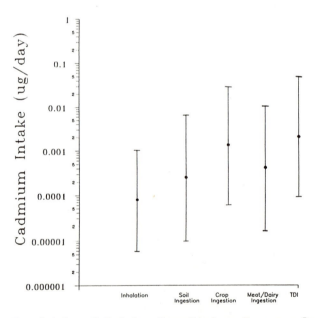

Figure 3.　Intake of Cadmium from Various Exposure Pathways.
Error Bars Represent 95% Confidence Intervals.
TDI Refers to Total Daily Intake from the
Food Chain Pathways.

Table 8. Contribution to Daily Intake of
         Cadmium by Vegetation Food Groups

Source	Percentage of the Total Intake
Garden fruits	32.3
Root vegetables	19.6
Potatoes	6.9
Leafy vegetables	4.7
Dried legumes	1.7
Grains	0.6
Fresh legumes	0.3

Table 9. Contribution to Daily Intake of
         Cadmium by Meat Food Groups

Source	Percentage of the Total Intake
Beef liver	12.3
Dairy products	3.7
Poultry	3.1
Beef	1.4
Pork	0.3
Eggs	0.2
Lamb	<0.1

Table 10. Distribution of Daily Cadmium Intake Values
          (ug/day)

Geometric Mean	2.11E-3
Geometric Standard Deviation	0.771
Minimum	3.90E-5
Maximum	7.19E-2

depicts the 95 percent confidence intervals for the daily intake of cadmium from the various exposure pathways, with inhalation included for comparison purposes. The inhalation pathway was calculated by multiplying the concentration of the chemical in air (ug/m^3) times the amount of air inhaled (20 m^3/day). Inhalation contributes only 3.7% to the total average individual (TAI) intake (inhalation plus food chain). The 95% confidence interval is less than a factor of 23 times the best estimate (i.e. geometric mean) of cadmium daily intake from the food chain. The range of uncertainty in TDI from the food chain across the interval is about 2.7 orders of magnitude.

Previous studies have shown that much of the overall uncertainty in estimates of human exposure to pollutants emitted from MWCs is attributable to uncertainty in the deposition component and the concentration value when modeling an "arbitrary" incinerator using cross-site data.[24,32] Hence, for comparison purposes model runs were made treating **Dy** and **Cy** as constants. Figure 4 depicts the 95 percent confidence interval for the total daily intake (TDI) of cadmium due to emissions of the contaminant from MWCs based on these point estimates. The 95% confidence interval is less than a factor of 2.1 times the best estimate (i.e. geometric mean) of cadmium daily intake from the food chain. The range of uncertainty in TDI from the food chain across the interval is approximately 4 times the lower bound.

Dioxin

Table 11 shows that the primary route of human exposure to dioxin from the food chain due to emissions from MWCs is through the consumption of contaminated meat and dairy products (88.6%). This pathway can be further subdivided into the soil-animal-human route (53.3%) and the plant-animal-human route (35.3%). Table 12 gives the contributions of each individual meat group, with beef (34.1%) and beef liver (23.8%) contributing the most.

While the soil ingestion by humans pathway contributes 3.3% to TDI, the consumption of contaminated vegetation contributes 8.1% (Table 11). The contaminated vegetation pathway can be subdivided into the air-to-plant transfer component (6.4%), the atmospheric deposition component (1.6%), and the root uptake component (0.1%). Table 13 shows that the primary vegetative group is garden fruits (4.4%).

A final goal of the uncertainty analysis was to determine a probability distribution for the daily human intake of dioxin due to emissions of the contaminant from MWCs. Figure 5 shows that these values fit a lognormal distribution, and their statistical summary is listed in Table 14.

Figure 6 depicts the 95 percent confidence intervals for the daily intake of dioxin from the various exposure pathways.

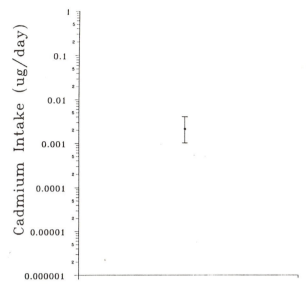

Figure 4.   95% Confidence Interval for TDI of Cadmium from the
Food Chain, using point estimates for Dy and Cy.

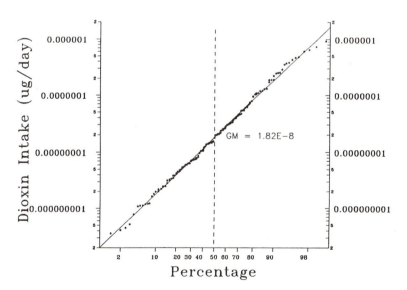

Figure 5.   Predicted Dioxin Daily Intake Values

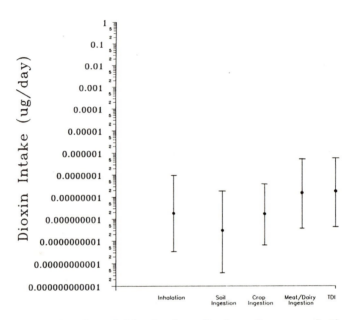

Figure 6.   Intake of Dioxin from Various Exposure Pathways.
Error Bars Represent 95% Confidence Intervals.
TDI Refers to Total Daily Intake from the
Food Chain Pathways.

Table 11. Breakdown of Exposure Pathways for Dioxin
           (% contributions to average daily intake)

---

1. Direct ingestion of contaminated soil :   3.3
2. Consumption of contaminated vegetation:

        Atmospheric deposition component   1.6
        Root Uptake component   0.1
        Air-to-plant transfer component   6.4

3. Consumption of contaminated meat and
    dairy products:

        Soil-Animal-Human Route   53.3
        Plant-Animal-Human Route   35.3

Table 12. Contribution to Daily Intake of
           Dioxin by Meat Food Groups

Source	Percentage of the Total Intake
Beef	34.1
Beef liver	23.8
Poultry	14.4
Dairy products	8.4
Eggs	6.3
Pork	1.6
Lamb	<0.1

Table 13. Contribution to Daily Intake of
           Dioxin by Vegetation Food Groups

Source	Percentage of the Total Intake
Garden fruits	4.4
Leafy vegetables	2.0
Dried legumes	1.3
Fresh legumes	0.3
Potatoes	0.1
Grains	<0.1
Root vegetables	<0.1

Table 14. Distribution of Daily Dioxin Intake Values
                        ($\underline{u}$g/day)

Geometric Mean	1.82E-8
Geometric Standard Deviation	0.782
Minimum	2.86E-10
Maximum	9.25E-7

An inhalation pathway is included for comparison purposes and shows that, in terms of the TAI (inhalation plus food chain), inhalation contributes 9.1%. The 95% confidence interval for TDI of dioxin from the food chain is approximately a factor of 40 times the best estimate. The estimates of dioxin intake across this interval range almost 3.2 orders of magnitude.

Model runs were also made which treated **Dy** and **Cy** as point estimates for comparison purposes. Figure 7 depicts the 95 percent confidence interval for the TDI of dioxin due to emissions of the pollutant from MWCs based on these estimates, which reflect no variability in the deposition rate and the concentration value. The 95% confidence interval is less than a factor of 1.6 times the best estimate of dioxin daily intake from the food chain. The range of uncertainty across the interval is approximately 2.2 times the lower bound.

CONCLUSIONS

The food chain is a primary pathway of human exposure to contaminants released by MWCs.[2,28] An uncertainty analysis was performed on the TFC Model to determine the extent of this exposure for two chemicals, cadmium and dioxin, known to be emitted from MWCs. Three main pathways of human exposure via the food chain were analyzed: (1) direct ingestion of contaminated soil; (2) consumption of contaminated vegetation; and (3) consumption of contaminated meat and dairy products. Three routes of food and forage crop contamination were considered: (1) atmospheric deposition; (2) uptake from the soil; and (3) air-to-plant transfer. The analysis showed that cadmium intake is due mainly to human consumption of contaminated plants, especially garden fruits and root vegetables. The main pathway of vegetation contamination is via root uptake. Dioxin intake occurs primarily through the ingestion of contaminated meat and dairy products, especially beef.

Based on deposition rates and concentration values for an "arbitrary" incinerator using cross-site data, distributions of cadmium and dioxin daily intake values were determined. Cadmium values ranged from 3.90E-5 to 7.19E-2 (ug/day), having a geometric mean of 2.11E-3. Also, the 95% confidence interval was less than a factor of 23 times the best estimate for cadmium exposure. Dioxin values varied from 2.86E-10 to 9.25E-7 (ug/day) and had a geometric mean of 1.82E-8, with the 95% confidence interval approximately a factor of 40 times the best estimate for dioxin total intake. Inhalation was shown to be a minor source of cadmium and dioxin intake in comparison with the contributions of the food chain pathways.

When distributions of total daily intake were determined based on point estimates for the deposition rate, **Dy**, and concentration value, **Cy**, uncertainty in intake estimates dramatically declined. These results showed that the 95% confidence intervals for adult total daily intake of cadmium and dioxin were less than a factor of 2.1 times the best estimate for both the metal and organic compound. Also, the total uncertainty in the daily intake estimates for both chemicals from all pathways was less than 0.5 orders of magnitude. These results support previous findings, indicating that only a small portion of the variability in human intake estimates is attributable to uncertainty in the Terrestrial Food Chain (TFC) Model.[32]

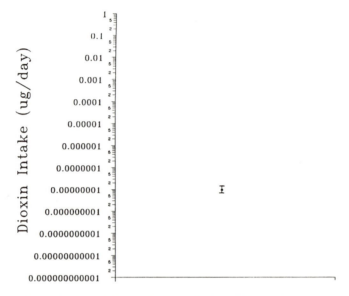

Figure 7.   95% Confidence Interval for TDI of Dioxin from the Food Chain, using point estimates for Dy and Cy.

Thus, the comparison of these two sets of results combined with the findings of the sensitivity analysis as listed in Table 1 strongly suggest that much of the overall uncertainty in estimates of human exposure to pollutants emitted from MWCs is attributable to uncertainty in the deposition rate and the concentration value when modeling an "arbitrary" incinerator using cross-site data. When determining human daily intake estimates based on a specific site in which deposition rates and concentration values experience limited variation, uncertainty in estimates of human exposure will likely be reduced,[33] with only a small degree of variability attributable to uncertainty in the TFC Model.

REFERENCES

1. Jones, K. H., Walsh, J. F., Siebert, P. C., and Alston,
        D. R., A joint probability model for predicting
        cross-sectional risks for MSW incinerator emissions
        of concern, in Proceedings of the International
        Conference on Municipal Waste Combustion, Hollywood,
        Florida, 1989.
2. Travis, C. C., and Hattemer-Frey, H. A., Human exposure
        to 2,3,7,8-TCDD, Chemosphere, 17, 277, 1987.
3. Travis, C. C., and Arms, A. D., The food chain as a
        source of toxic chemical exposure, in Toxic
        Chemicals, Health and the Environment, Lave, L. B.,
        and Upton, A. C., Eds., Plenum Press, New York,
        1987, p. 75.
4. Hattemer-Frey, H. A., and Travis, C. C.,
        Pentachlorophenol: environmental partitioning and
        human exposure, Arch. Environ. Contam. Toxicol., 18,
        482, 1989.
5. Travis, C. C., Holton, G. A., Etnier, E. L., Cook, C.,
        O'Donnell, F. R., Hetrick, D. M., and Dixon, E.,
        Assessment of inhalation and ingestion population
        exposures from incinerated hazardous wastes, 12,
        533, 1986.
6. Hoffman, F. O., and Gardner, R. H., Evaluation of
        uncertainties in radiological assessment models, in
        Radiological Assessment: A Textbook on
        Environmental Dose Analysis, Till, J. E., and Meyer,
        H. R., Eds., NUREG/CR-3332, 1983, chap. 11.
7. Ronen, Y., The role of uncertainties, in Uncertainty
        Analysis, Ronen, Y., Ed., CRC Press, Boca Raton,
        1988, chap. 1.
8. Gardner, R. H., Rojder, B., and Bergstrom, U., PRISM: a
        systematic method for determining the effect of
        parameter uncertainties on model predictions,
        Studvisk Energiteknik AB Report STUDSVIK/NW-83/555,
        Nykoping, Sweden, 1983.
9. Ronen, Y., Sensitivity and uncertainty analysis using a
        statistical sample of input values, in Uncertainty
        Analysis, Ronen, Y., Ed., CRC Press, Boca Raton,
        1988, chap. 4.
10. Gardner, R. H. personal communication, 1989.
11. Ronen, Y., Uncertainty analysis based on sensitivity
        analysis, in Uncertainty Analysis, Ronen, Y., Ed.,
        CRC Press, Boca Raton, 1988, chap 2.
12. Hoffman, F. O., and Baes, C. F., A statistical analysis
        of selected parameters for predicting food chain
        transport and internal dose of radionuclides, Report
        ORNL/NUREG/TM-882, Oak Ridge National Laboratory,
        1979.
13. Yang, Y., and Nelson, C. B., An estimation of daily food
        usage factors for assessing radionuclide intakes in
        the U.S. population, Health Physics, 50, 245, 1986.
14. Baes, C. F., Sharp, R. D., Sjoreen, R. W., A review and
        analysis of parameters for assessing transport of
        environmentally released radionuclides through

agriculture, Report ORNL-5786, Oak Ridge National Laboratory, 1984.

15. Watt, B. K., and Merrill, A. L., Composition of foods, Agricultural Handbook No. 8, U.S. Department of Agriculture, Washington, D.C., 1963.

16. Eschenroeder, A., Guldberg, P., Hahn, J., Kellermeyer, D., Smith, A., Wolff, S., and Ziemer, S., An analysis of health risks from the Irwindale resource recovery facility, California Energy Commission, Sacramento, 1986.

17. Lagoy, P. K., Estimated soil ingestion rates for use in risk assessment, Risk Analysis, 7, 355, 1987.

18. Kimbrough, R. D., et al., Health implications of 2.3,7,8-tetrachloro-dibenzodioxin (TCDD) contamination of residential soil, Journal of Toxicology and Environmental Health, 14, 47, 1984.

19. U.S. Environmental Protection Agency, Methodology for the assessment of health risks associated with municipal waste combustor emissions, Report EPA-530/SW-87-021g, U.S. Environmental Protection Agency, Cincinnati, Ohio, 1986.

20. U.S. Environmental Protection Agency, Dietary consumption distributions of selected food groups for the U.S. population, Office of Toxic Substances, Washington, D.C., 1980.

21. U.S. Bureau of the Census, Statistical abstract of the United States: 1987, 107th edition, Washington, D.C., 1986.

22. Shor, R. W., Baes III, C. F., and Sharp, R., Agricultural production in the United States by county: a compilation of information from the 1974 census of agriculture for use in terrestrial food-chain transport and assessment models, Report ORNL-5768, Oak Ridge National Laboratory, 1982.

23. Travis, C. C., and Arms, A. D., Bioconcentration of organics in beef, milk, and vegetation, Environ. Sci. Technol., 22, 271, 1988.

24. McKone, T. E., and Ryan, P. B., Human exposures to chemicals through food chains, Environ. Sci. Technol., 23, 1154, 1989.

25. Ng, Y., Transfer coefficients for assessing the dose from radionuclides in meat and eggs, Report NUREG/CR02967, U.S. Government Printing Office, Washington, D.C., 1980.

26. Fries, G. F., Assessment of potential residues in food derived from animals exposed to TCDD-contaminated soil, Chemosphere, 16, 2123, 1987.

27. Risk assessment study of the Dickerson site, Maryland Power Plant and Review Division, Montgomery Co., Maryland, 1988.

28. Hattemer-Frey, H. A., and Travis, C. C., An overview of food chain impacts from municipal waste combustion, in Proceedings of the U.S. EPA/ORNL Workshop on Risk Assessment for Municipal Waste Combustion: Deposition, Food Chain Impacts, Uncertainty, and Research Needs, Plenum Publishing Corporation, New York, 1989.

29. Lowe, J. A., Dietrick, R. W., Alberts, M. T., Health risk assessments for waste-to-energy projects in California, in <u>Municipal Waste Incineration and Human Health</u>, Travis, C. C., and Hattemer-Frey, H. A., Eds., CRC Press, Boca Raton, 1989.

30. U. S. Environmental Protection Agency, Industrial source complex (ISC) dispersion model user's guide, Vol. I, Report EPA-450/4-79-030, H. E. Cramer Company, Salt Lake City, Utah, 1979.

31. Miller, C., and Hoffman, F., An examination of the environmental half-time for radionuclides deposited on vegetation, <u>Health Physics</u>, 45, 731, 1983.

32. Belcher, G. D., and Travis, C. C., The food chain as a source of human exposure from municipal waste combustion, in <u>Proceedings of the International Conference on Municipal Waste Combustion</u>, Hollywood, Florida, 1989.

33. California Energy Commission, Exposure estimation, San Diego Energy Recovery (SANDER) Project, Appendix C, Part 2, Public Health Report 06-24-88/4254e, 1988.

34. Heffron, C. L., Cadmium and zinc in growing sheep fed silage corn grown on municipal sludge amended soil, <u>J. Agric. Food Chem.</u>, 18, 58, 1980.

35. U.S. Environmental Protection Agency, Environmental profiles and hazard indices for constituents of municipal sludge: cadmium, Office of Water Regulations and Standards, Washington, D.C., 1985.

36. Health risk assessment for a resource recovery facility in Montgomery County, Maryland, submitted to the Montgomery County government, prepared by Weston, Inc., 1988.

37. San Diego energy recovery (SANDER) project, revised health risk assessment, submitted to the California Energy Commission, prepared by Signal Environmental Systems, Inc., 1987.

38. Risk assessment for the proposed trash-to-steam municipal solid waste incinerator at the U.S. naval base in Philadelphia, Pennsylvania, submitted to the Philadelphia Public Health Advisory Commission, prepared by Clement Associates, Inc., 1986.

39. Stanislaus waste-to-energy facility health risk assessment, submitted to the Stanislaus County Air Pollution Control District, prepared by the Stanislaus Waste Energy Co., 1985.

40. Final environmental impact report of the Los Angeles City energy recovery (LANCER) project, submitted to the Department of Public Works, Bureau of Sanitation, prepared by Cooper Engineers, Inc., 1985.

41. Health risk assessment for the Brooklyn navy yard resource recovery facility, prepared by Dr. Allan H. Smith, 1988.

42. Bloomington incinerator project risk assessment, prepared by the Westinghouse Electric Corporation, Waste Technology Services Division, 1986.

43. Assessment of health risks associated with trace elements and organic emissions from the proposed SPADRA refuse-to-energy project, Pomona, California, HDR Techserv, Inc, Santa Barbara, CA, 1987.

44. Draft site and technology specific impact addendum to the generic environmental impact statement for the North Hempstead solid waste management facility project, health risk assessment, prepared by Malcolm Pirnie, 1987.

45. Garten, C. T., and Trabalka, J. R., Evaluation of models for predicting terrestrial food chain behavior of xenobiotics, Environ. Sci. Technol., 17, 590, 1983.

46. Washburn, S. T., The accumulation of chlorinated dibenzo-p-dioxins and dibenzofurans in beef and milk, in Municipal Waste Incineration and Human Health, Travis, C. C., and Hattemer-Frey, H. A., Eds., CRC Press, Boca Raton, 1989.

47. Travis, C. C., Yambert, M. W., and Arms, A. D., Food chain exposure from municipal waste incineration, Oak Ridge National Laboratory, submitted to the Risk Assessment Study of the Dickerson Site, 1988.

48. Isensee, A. R., and Jones, G. E., Absorption and translocation of root and foliage applied 2,4-dichlorophenol, 2,7-dichlorodibenzo-p-dioxin, and 2,3,7,8-tetrachlorodibenzo-p-dioxin, J. Agr. Food Chem., 19, 1210, 1971.

49. U.S. Environmental Protection Agency, Superfund public health evaluation manual, Office of Emergency Response and Remedial Response, EPA/540/1-86/060, 1986.

50. McKone, T. E., The use of environmental health-risk analysis for managing toxic substances, presented at the 78th annual meeting of the APCA, Detroit, MI, 1985.

51. Crosby, D., The degradation and disposal of chlorinated dioxins, in Dioxins in the Environment, Kamrin, M. A., and Rodgers, P. W., Eds., Hemisphere Press, Washington, D.C., 1985.

52. Briggs, G., Theoretical and experimental relationships between soil adsorption, octanol-water partition coefficients, water solubilities, bioconcentration factors, and the parachor, J. Agric. Food Chem., 29, 1050, 1981.

53. U.S. Environmental Protection Agency, Municipal waste combustion study, emission data base for municipal waste combustors, Offices of Solid Waste and Emergency Response, Air and Radiation, and Research and Development, EPA/530-SW-87-021b, 1987.

54. U.S. Environmental Protection Agency, Preliminary risk calculations for deposited contaminants from municipal waste combustor emissions: polychlorinated dibenzo-p-dioxins (PCDD) and polychlorinated dibenzofurans (PCDF), Office of Health and Environmental Assessment, Environmental Criteria and Assessment Office, Cincinnati, 1987, (Template).

55. U.S. Environmental Protection Agency, Environmental profiles and hazard indices for constituents of municipal sludge: chlorinated dioxins, Office of Water Regulations and Standards, Washington, D.C., 1985.

56. U.S. Environmental Protection Agency, Municipal waste
     combustion study, characterization of the municipal
     waste combustion industry, Offices of Solid Waste
     and Emergency Response, Air and Radiation, and
     Research and Development, EPA/530-SW-87-021h, 1987.
57. Milliken waste-to-energy project health risk assessment,
     submitted to the South Coast Air Quality Management
     District, prepared by Radian Corporation, 1987.

ASSESSING MULTIPLE PATHWAY EXPOSURES: VARIABILITY, UNCERTAINTY, AND IGNORANCE

Thomas E. McKone
Lawrence Livermore National Laboratory
Livermore, California

ABSTRACT

Human populations contact environmental pollutants through food, water, and air in varying amounts each day throughout a lifetime. Thus, a realistic strategy for managing the potential health risks of municipal incinerator emissions requires a comprehensive approach with adequate attention to uncertainties. Using contaminant transfers from air to food as a case study, this paper considers two important issues in exposure assessment–completeness of the exposure model and the treatment of uncertainty in exposure estimates. This case study is used to distinguish between variability (inherent randomness in data), ignorance (incomplete data and/or lack of scientific understanding) and uncertainty (the variance in exposure estimates attributable to the combination of variability and ignorance). For the air/food pathways, I explore the use of pathway exposure factors (PEFs) that combine information on environmental partitioning ("fugacity," biotransfer factors, deposition, etc.) with data on human diet, behavior patterns, and physiology into a numerical expression that links ambient air concentrations in mg/m^3 into daily exposure in mg/kg-d. Following EPA protocol, exposure expresses human contact with contaminants through the lungs, the gut wall, and skin surface. I describe and assess the uncertainty for exposure estimates of incinerator emissions through the air/milk and air/meat pathways. I consider the advantages and disadvantages of various methods for propagating and analyzing uncertainties.

Municipal Waste Incineration Risk Assessment
Edited by C.C. Travis, Plenum Press, New York, 1991

INTRODUCTION

Based on a definition of exposure as "the contact with a chemical or physical agent" (EPA, 1987), human exposure assessments translate contaminant sources into quantitative estimates of the amount of contaminant that passes through the lungs, across the lining of the GI tract, and through the skin of individuals within a specified population. In current practice, exposure assessments rely either on the measurement of a contaminant or its surrogates or on concentration estimates from contaminant-transport models. Often implicit in these methods is the assumption that exposure can be linked by simple parameters to ambient concentrations in air, water, and soil. However, assessments that include time and activity patterns and microenvironmental data, such as the TEAM (Wallace, 1986) and Harvard Health (six-cities) studies (Spengler et al., 1981), reveal the importance of providing a complete picture of human exposure. An exposure assessment may be most valuable when it provides a comprehensive view of exposure routes and identifies major sources of uncertainty and what impact this will have on the decision-making process. Yet, the common practice in exposure evaluations has often been to use single exposure routes and mean or point estimates for most parameters.

Overview

Human exposures to ambient airborne pollutants is an area where uncertainty is particularly important. This uncertainty can be divided into variability and ignorance. Variability is the result of inherent randomness in data; ignorance results from incomplete data and/or lack of scientific understanding. Consider for example the chemical 2,3,7,8-tetrachlorodibenzo-p-dioxin (TCDD), which is found in ambient air attached to suspended dust or combustion particles. This contaminant can be found in the air as a result of hazardous-waste incineration or as a result of land disposal of municipal or industrial wastes. Inhalation is traditionally the pathway given the most attention even though food-chain pathways, which are less well understood, might be the dominant contributor to total exposure.

In this paper I use simple steady-state air/plant/food-ingestion models to examine uncertainties in the daily human exposure associated with transfer of chemicals from air emissions to humans. For the purposes of this paper, I limit attention to meat and milk pathways and air as a transport medium. This should not be interpreted as implying that transfers from other environmental media through alternate pathways (i.e., fruit, vegetable, grain ingestion; inhalation; and dermal absorption) are unimportant or more precisely understood. McKone

(1988) and McKone and Ryan (1989) describe a more comprehensive matrix of environmental media and exposure pathways and the PEFs that link them. Using these pathways as case studies, I will consider the approach used to assess risks associated with incinerator emissions, and address the factors that limit our ability to predict precisely human exposure through that pathway. In particular, I will consider the limits on precision imposed by uncertainties in deposition and biotransfer factors; methods for assessing this lack of precision or uncertainty; and strategies for reducing uncertainty through enhancement of data or model improvement. One of the important issues in this analysis is the relative contribution to uncertainty from inherent variabilities in natural processes, data limitations, and the incomplete nature of our models. We might be able to increase the complexity of our models and thus improve their credibility, but unless we concurrently improve the precision of the inputs we will not increase the precision of the outputs.

Exposure and Risk

A risk assessment by its very nature involves uncertain consequences and thus one useful output is a probability distribution characterized by expectation (mean) and spread (variance). One can represent the overall process of predicting the population risk resulting from a contaminant released at a source S (in kg/y) as the function of five factors:

$$R = N \times f(S, E(S), D(E), \beta) \tag{1}$$

where R is the risk expressed as the annual incidence of disease associated with the source S in a population of size N; E is the exposure function, which converts the source into a lifetime equivalent contact in the population; D is the dose factor, the fraction of contaminant delivered to the organism after contact; and β is the potency (i.e., risk per unit dose rate) associated with the delivered dose. In actual practice, the process of estimating exposure and risk is more complex and includes temporal and/or spatial relations and functional dependencies among the source, exposure, dose and the incidence of specified health effects.

In a predictive model, the exposure function E is divided into a series of terms that relate sources to environmental concentrations and the concentrations to human contact:

$$E(S) = \sum_i \sum_j C_i(S) \times F_{ij} \qquad \begin{aligned} &i=\text{air, soil, water} \\ &j=\text{inhalation, ingestion, dermal} \end{aligned} \tag{2}$$

where $C_i(S)$ is the concentration of a contaminant in environmental compartment i associated with the source S and F_{ij} is the pathway exposure factor (PEF) that relates this concentration to a level of human contact through pathway j. Figure 1 illustrates the relation between environmental transport and human exposure in a comprehensive model.

The movement of contaminants among environmental media is affected by physical, chemical, and biological processes, including leaching from soil to ground water, biotransformation, volatilization from water or soil, and deposition from air to land and water. The nature of these processes requires a multimedia approach. In contrast to the single medium paradigm for assessing exposure, a multimedia approach locates all points of release to the environment; characterizes mass balance relationships (e.g., between sources and sinks in the environment); traces contaminants through the entire environmental system, observing and recording changes in form as they occur; and identifies where in this chain of events control efforts would be most appropriate.

In order to link contaminant concentrations in water, air, or soil, I use pathway exposure factors (PEFs). The PEF combines a mix of information on environmental partitioning, human physiology, and behavior patterns into a numerical term that converts concentrations (in mg/L water, mg/m^3 in air and mg/kg soil) into a daily exposure in mg/kg-d for a specific route such as inhalation, ingestion, or dermal uptake. For example, the PEF for exposure of a 70-kg adult to a contaminant contained in 2 L of water consumed daily is calculated as 2 L divided by 70 kg or 0.0286 (mg/kg-d)/(mg/L).

MODELLING THE EXPOSURE PATHWAYS FROM AIR PARTICLES TO MEAT AND MILK

The PEFs F_{pt} and F_{pk} account for human contaminant exposures attributable to meat- and milk-producing cattle as a result of direct inhalation of particles and ingestion of particles deposited onto pasture. The expressions that link the atmospheric concentration C_p and the two PEFs to human exposure (in mg/kg-d) through meat, $e_t(C_p)$, and dairy products, $e_k(C_p)$, are as follows:

$$e_t(C_p) = F_{pt} C_p \qquad [3]$$

$$e_k(C_p) = F_{pk} C_p \qquad [4]$$

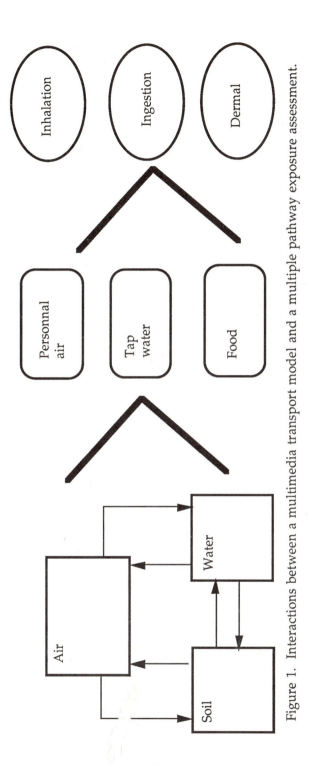

Figure 1. Interactions between a multimedia transport model and a multiple pathway exposure assessment.

221

The PEFs in Equations 3 and 4 are related to human ingestion of meat and dairy products; cattle properties; deposition factors; and meat and dairy-product partition factors. The factors F_{pk} and F_{pt} are used to relate ingestion exposure solely to concentration in the air and not to the air source term. This means that these PEFs do not account for deposition/resuspension processes, which are assumed to be included in the estimation or measurement of the concentration C_p. In order to calculate the transfer of atmospheric particles from atmosphere to milk and meat; we must consider material inhaled by cattle and the balance between material that deposits on the exposed and edible portion of pasture crops, is removed by weathering and senescence, and ingested by milk- and meat-producing cattle. Using this approach, the PEF for transfer from atmosphere to milk is set up as the ratio of intake to atmospheric concentration,

$$F_{pk} = \text{(daily contaminant intake by humans)}/C_p \qquad [5]$$

and the steady-state concentration in pasture is obtained under the assumption that gains equal losses,

$$\frac{dCv}{dt} = 0 = V_{dp}\, C_p - M_p\, R_v\, C_v \qquad [6]$$

where V_{dp} is the deposition factor that has units of m/d and relates the deposition due to wet and dry processes onto the surface of pasture to the concentration C_p (mg/m^3) in the air of particle-bound contaminants; M_p is the annual average inventory of pasture crops per unit area, kg(dry mass)/m^2; R_v is the rate constant for the removal of chemicals from vegetation surfaces as a result of weathering and senescence, 1/day; and C_v is the resulting contaminant concentration in pasture vegetation, mg/kg (dry mass).

The resulting intake of contaminants by beef and dairy cattle are given by

$$\text{Intake(beef cattle)} = I_{nc}\, C_p + I_{vbc}\, C_v, \text{ and} \qquad [7]$$

$$\text{Intake(dairy cattle)} = I_{nc}\, C_p + I_{vdc}\, C_v, \qquad [8]$$

where, I_{nc} is the inhalation rate for cattle (beef and dairy), m^3/d; I_{vbc} is the ingestion rate of pasture grasses by beef cattle, kg(dry mass)/d; and I_{vdc} is the ingestion rate of pasture grasses by dairy cattle, kg(dry mass)/d. By combining Equations 5 through 8 and including human ingestion data together with local consumption factors, I obtain the following expressions for the meat and milk PEFs,

$$F_{pt} = I_t \left[I_{nc} + (I_{vbc} V_{dp})/(M_p R_v) \right] f_t B_t \qquad\qquad [9]$$

$$F_{pk} = I_k \left[I_{nc} + (I_{vdc} V_{dp})/(M_p R_v) \right] f_k B_k \qquad\qquad [10]$$

where, I_t is human intake per unit body weight of meat, kg(fresh mass)/kg-d; I_k is human intake per unit body weight of dairy products, kg/kg-d; f_t, f_k are, respectively, fractions of the target population's meat and dairy products that come from the area with concentration, C_p, no units; B_t is the biotransfer factor from cattle intake to meat, which is the steady-state contaminant concentration in meat divided by the animal's daily contaminant intake, (mg/kg)/(mg/d); and B_k is the biotransfer factor from cattle intake to dairy products, which is the steady-state contaminant concentration in dairy products divided by the animal's daily contaminant intake, (mg/L)/(mg/d).

MODEL AND INPUT UNCERTAINTIES

In this section, I review the type of data available to estimate the parameters in Equations 9 and 10 and discuss the sources and magnitudes of uncertainties associated with these parameters.

Deposition
        The deposition factor is the ratio of deposition rate on vegetation in mg/m²-d to the air concentration in mg/m³. The deposition factor V_{dp} includes both wet and dry deposition processes,

$$V_{dp} = v_g + b R W_p \qquad\qquad [11]$$

where, v_g is the dry deposition velocity onto vegetation, m/d; b is the fraction of material retained on vegetation from wet deposition; R is the annual average rainfall rate, m/d; and W_p is the volumetric washout factor for particles, no units. According to Peterson (1983), the retention factor b depends on the intensity of rainfall and is in the range 0.1 to 0.3. I assume an annual rainfall on the order of 1 m. Bidleman (1988) reports that W_p is highly variable and controlled by meteorological factors and particle size and that a large number of measurements of W_p for trace metals in air and rain had geometric means in the range 2×10^5 to 1×10^6 and geometric standard deviations in the range 2.2 to 3.2.
        Dry deposition velocities are influenced by numerous factors and there are a wide range of reported values, for example from 3 to 4900 m/d reported for deposition rates of particles from air to vegetation surfaces (Whicker and

Kirchner, 1987). Schroeder and Lane (1988) report that dry deposition velocities measured for gases span four orders of magnitude, from 0.002 cm/s (1.7 m/d) to 26 cm/s (22,000 m/d). They report deposition velocities measured for particles in the range from 0.001 cm/s (0.86 m/d) to 180 cm/s (155,000 m/d). For particles less than 5 μm, McMahon and Denison (1979) report deposition velocities in the range 0.003 to 1 cm/s (2.6 to 860 m/d). According to Whicker and Kirchner (1987), the fraction of deposited particles intercepted by vegetation can be calculated from the total deposition velocity v_t as

$$v_g = v_t \,[1\text{-}\exp(\text{-}\alpha M_v)] \tag{12}$$

where α is the foliar interception constant, m^2 per dry kg, and M_v is the dry mass inventory of vegetation per unit area, kg/m^2. Using a fixed foliar interception constant of 2.8 (Whicker and Kirchner, 1987) and an annual average dry-mass inventory of 0.6 kg/m^2 (Bowen, 1979), one can estimate that v_g is on the order of $0.8v_t$ for food and pasture crops.

Based on the observations above, I represent uncertainty in the deposition factor V_{dp} from air particles (less than 5 μm) to vegetation surfaces using a log-normal distribution having a geometric mean of 300 m/d and a geometric standard deviation of 3.0. It is important to note that uncertainty in the deposition factor comes from both the natural variability in the underlying processes and lack of complete information on the parameters in the model. To the extent that one can make chemical- and site-specific measurements to reduce the latter source of uncertainty, the overall uncertainty about deposition might be reduced. However, for many cases a risk assessment must be carried out before such measurements can be made.

Biotransfer

For organic chemicals the biotransfer coefficients B_t and B_k can be related to the fat-diet partition coefficient using the following expressions:

$$B_t = K_{fd} \times 0.4/I_{vbc} \tag{13}$$

$$B_k = K_{fd} \times 0.05/I_{vdc} \tag{14}$$

where K_{fd} is the fat/diet partition coefficient and expresses the ratio of contaminant concentration in animal fat to that in animal feed, (mg/kg)/[mg/kg(dry mass)] and 0.4 and 0.05 are the fractional fat content of, respectively, meat and dairy products (Layton et al., 1986). Variance in fat-content values is likely to be small relative to the variance in fat/diet partition coefficients. Kenaga (1980) has developed the following correlation for estimating fat/diet partition coefficients,

$$\text{Log } K_{fd} = 0.5 \text{ Log } K_{ow} - 3.457 \pm 2.0 \qquad (n = 23, r = 0.79) \tag{15}$$

in which K_{ow} is the octanol/water partition coefficient and the $\pm$ range reflects the 95% confidence bound for this estimate.

Travis and Arms (1988) have reviewed biotransfer factors for 36 organic chemicals in meat and 28 organic chemicals in milk. For these two pathways they have developed geometric mean regressions for the biotransfer factors in terms of K_{ow}. Their reported regressions are

$$\text{Log } B_t = \text{Log } K_{ow} - 7.6 \qquad (n = 36, r = 0.81) \text{ and} \tag{16}$$

$$\text{Log } B_k = \text{Log } K_{ow} - 8.1 \qquad (n = 28, r = 0.74). \tag{17}$$

Travis and Arms (1988) have not estimated the uncertainty in their correlation. Figure 2 shows the 36 data points used by Travis and Arms to correlate Log B_t with K_{ow} along with lines corresponding to a standard least squares fit through the data and a geometric least squares fit (as is used by Travis and Arms). Also shown in Figure 2 are the two curves that represent the 95% confidence interval about the standard least squares fit. These latter curves are obtained from a procedure described in Wonnacott and Wonnacott (1985). Based on this analysis, I estimate that the 95% prediction interval for both B_t and B_k based on equations 16 and 17 above is approximately two orders of magnitude and thus similar to the variability reported by Kenaga (1980).

The B_t and B_k values used in the example below for TCDD and taken from the Travis and Arms data (1988) are not estimated from the regression equations above but based on experimental data (Jensen and Hummel, 1982). In order to estimate the geometric standard deviation of biotransfer factors measured for TCDD, I assume that biotransfer experiments with organic compounds can be no more precise than those with nonessential radioactive elements, which can be measured with a geometric standard deviation of 3 (Ng, 1982). This is lower than the geometric standard deviation of 10 associated with the regression estimations in equations 16 and 17.

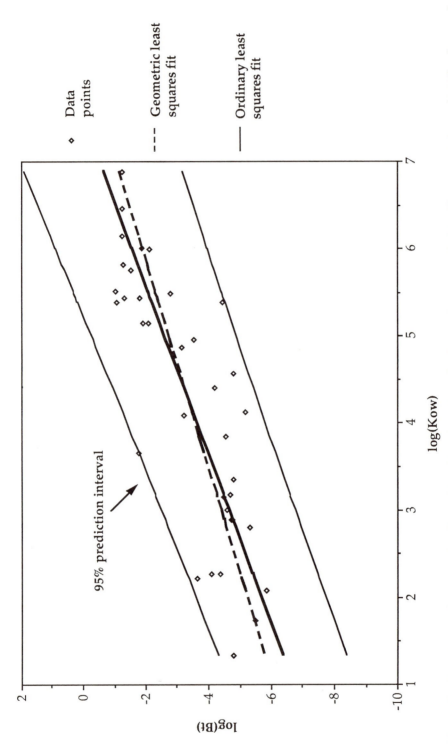

Figure 2. Correlations between log B_t and log K_{OW} based on 36 data points compiled by Travis and Arms (1988) who used these data to develop a geometric least squares fit of log B_t versus log K_{OW}; also shown is an ordinary least squares fit through the data points and the two curves that bound the 95% confidence interval about the ordinary least squares fit.

Human Ingestion Rates

Yang and Nelson (1986) analyzed statistically data from the 1977-1978 USDA Food Consumption Survey to estimate the daily intake by food group for various age categories and regions within the U.S. This survey provides a stratified sample of households in the 48 conterminous states and the District of Columbia for three days in each of four seasons from April 1977 to March 1978. Samples were classified according to the geographic region of the country; geographic divisions within these regions; and central city, suburban and nonmetropolitan populations. Data were collected for 30,770 individuals within 114 primary sampling units–approximately 270 individuals per sampling unit.

Data on the age distribution of and variance of body mass in human populations have been reviewed and published by the ICRP (1975). Based on these data, one can determine that the distribution of body mass among adult males and females (ages 15 and up) has an arithmetic mean of 66 kg and standard deviation of 14 kg and the distribution of body mass among male and female children (ages 0 to 15) has an arithmetic mean of 27 kg and standard deviation of 14 kg.

Based on data compiled by Yang and Nelson (1986), McKone and Ryan (1989) have determined that the daily intakes of milk by children and adults in the U.S. have mean and standard deviations of 0.39 ± 0.2 and 0.22 ± 0.08 respectively and daily intakes of meat are 0.12 ± 0.05 and 0.19 ± 0.09 respectively. Combining this information with the distributions of body weight given above, I estimate that the intake of milk and meat per unit body weight is as listed in Table I. The variance in food intake per unit body weight is calculated using a covariance that is derived from the assumption that food intake scales with body weight to the two-thirds power.

Data for Cattle

In order to calculate human exposures to chemicals in dairy products and meat we need estimates of inhalation (I_{nc}) and ingestion rates for both beef (I_{vbc}) and dairy cattle (I_{vdc}). Table I lists the ranges of cattle properties that were obtained by Ryan and McKone (1989) from eight papers, which addressed the transfer of contaminants from air, vegetation, or soil to beef and dairy cattle. Because there is no basis by which to rank the quality of information in these papers, they are given equal weight when developing input data distributions

Table I. Type of distributions and distribution moments used for the input parameters.

Parameter	Expected value	Geometric mean	GSD[a]	Distribution type
I_t (child)[b]	0.0044	0.0044	1.1	log normal
I_t (adult)[b]	0.0030	0.0029	1.2	log normal
I_k (child)[c]	0.015	0.014	1.2	log normal
I_k (adult)[c]	0.0033	0.0033	1.1	log normal
V_{dp}	500	300	3.0	log normal
B_t (TCDD)	0.10	0.055	3.0	log normal
B_k (TCDD)	0.02	0.010	3.0	log normal

Parameter	Expected value	Lower bound	Upper bound	Distribution type
M_p	0.39	0.1	1.0	log uniform
R_v	0.039	0.01	0.1	log uniform
I_{nc}	120	63	177	uniform
I_{vbc}	12	4.0	20.	uniform
I_{vdc}	17	11	23	uniform

[a] GSD = geometric standard deviation
[b] I_t refers to milk ingestion by humans per unit body weight (kg/kg-d)
[c] I_k refers to meat ingestion by humans per unit body weight (kg/kg-d)

and this results in the uniform distributions. The uniform distributions also give us a good fit to the upper and lower bounds, the mean, and the standard deviation of the reference data.

Pasture Crop Inventory and Removal Rate Constant

Whicker and Kirchner (1987) and Bowen (1979) report the inventory of standing and mature biomass in agricultural landscapes to be on the order of 0.63 kg(dry mass)/m^2 or 3.0 kg(fresh mass)/m^2. Layton et al. (1986) suggest that the variation in standing biomass is within a factor of three higher or lower than this value. Whicker and Kirchner (1987) provide weathering data with which I estimate that the rate constant R_v is on the order of 0.03 day^{-1} with a range of 0.01

to 1.0 day^{-1}. Because these parameters have values that vary by a factor of ten or more and for which there is no other information indicating the character of the distribution, I represent the uncertainty in them using log-uniform distributions.

METHODS FOR ASSESSING AND EVALUATING UNCERTAINTY

At best, mathematical models only approximate real systems; and therefore their predictions are inherently uncertain. One of the key questions that must be addressed in an uncertainty analysis is -- how large is the uncertainty in the model predictions? This question can be answered by considering the uncertainty or variability of each parameter that is used in a predictive model and by propagating this uncertainty through to the model predictions. Two main classes of propagations methods are often used to relate parameter uncertainty to output uncertainty. These are analytical methods and numerical methods.

<u>Analytic Methods</u>

Analytical methods for parameter uncertainty analysis are capable of quickly providing necessary information on the probability distribution of model predictions without assistance of a computer. However, such methods are only practical for simple models. Sometimes models structure can be simplified in order to make analytical methods applicable. Analytical methods require that the relationship between the model prediction and the set of uncertain parameters be expressed as an algebraic equation. Among the most commonly used analytical methods are variance propagation and moment matching.

Variance propagation is easily applied to simple additive or multiplicative models and for models that can be simplified to a chain of additive or multiplicative terms. For example consider two parameters X and Y with standard deviations s_x and s_y. The variance s_z^2 in the sum $Z = aX \pm bY$ is calculated as

$$s_z^2 = a^2 s_x^2 + b^2 s_y^2 \pm 2ab\, s_{xy}^2, \qquad [18]$$

and the variance s_z^2 in the product $Z = X\,Y$ is calculated as

$$(s_z/z)^2 = (s_x/X)^2 + (s_y/Y)^2 \pm 2s_{xy}^2/(X\,Y), \qquad [19]$$

where s_{xy}^2 is the covariance between the variable x and y.

Moment matching permits the derivation of subjective confidence limits and intervals by identifying a distribution function with the same mean, variance, and third and fourth central moments as the probability density function (PDF) of the model predictions. Thus moment matching requires that the above four moments of the unknown PDF are obtained from given moments of the uncertain input parameters. Additional information on moment matching can be found in Hahn and Shapiro (1967).

Numerical Methods

Consider a model with m uncertain input parameters. Each of the input parameters can be represented by a probability density function that defines both the range of values that parameter can take on and the likelihood that the parameter has a value in any subinterval of that range. In order to make a prediction using this model, one needs a set of m values (or an ordered m-tuple) to define the input parameters. Numerical methods for propagating uncertainty simply involve the selection of several m-tuples and the computation of the corresponding predicted values with the prediction model. Numerical methods are applicable even when model predictions cannot be provided as a simple algebraic expression of the input parameters.

Monte Carlo Methods. These methods use random sampling to select each member of the m-tuple set. The most common Monte Carlo approach involves the use of simple random sampling. A random sample of each parameter X is selected by sampling randomly on the interval [0,1] and value of a parameter t. The selected value of t is transformed into value x of parameter X by choosing the smallest value of x such that the probability that the value x exceeds the actual value of the parameter X is less than or equal to t [i.e. $P(x>X) \leq t$]. If n values of X are needed, this procedure is repeated n times and each time the procedure is repeated, every value from the range [0,1] has an equal chance of being selected. In the special case where there are no correlations between the input parameters, each of the m parameter values in an m-tuple can be selected independently.

Latin Hypercube Sampling . This is a Monte Carlo method that uses stratified random sampling to select each member of an m-tuple set. If n values of the parameter X are needed, the interval [0,1] is divided into n intervals of equal length. One of these subintervals is selected at random and designated i1. A value t1 is selected at random from i1. This value of t1 is transformed into a value x1 following the Monte Carlo process described above. The overall process is repeated for each of the n intervals. Whereas for simple random sampling it is often a matter of chance how evenly the n selected values cover the range of parameter X, latin hypercube sampling places restrictions on possible unevenness. Additional information on latin hypercube sampling is available in Iman and Shortencarier (1984)

UNCERTAINTY IN TCDD EXPOSURES THROUGH MILK AND MEAT

I calculate here probability distributions for the exposures attributable to a unit concentration (1 pg/m^3) of 2,3,7,8-TCDD attached to airborne particles. Table I summarizes the mean, standard deviation, and type of distribution used to represent each of the parameters in the equations that define PEFs. I use the log-normal distributions to represent the intake-per-unit-body-weight data because these parameters are positive and based on the quotient of uncertain data. Log-uniform distributions are used to represent the uncertainty in parameters that are expected to have a value that varies by a factor of ten or more and for which there is no other information indicating the character of the distribution. I found that the data on the properties of cattle are best fit by uniform distributions. Following the suggestion of Ng (1982), I represent uncertainty in biotransfer coefficients using log-normal distributions.

Distributions in Table I are used in a Monte Carlo simulation to produce the distributions of PEFs and exposures. The Monte Carlo simulations are carried out with latin-hypercube sampling using the LHS program (Iman and Shortencarier, 1984) and the SIMSYS program developed by Ryan and Letz (1984-1988). The LHS program is used to produce 1000 stratified sample values for each parameter used in the defining equation for a given PEF. The restricted pairing procedure of the LHS program is used to minimize spurious correlations between input parameters. The SIMSYS program is used to read the LHS-generated data, construct the distribution of PEFs and exposure estimates, and to evaluate the properties of the output distributions.

Table II lists the outcome of this simulation by age group and pathway for exposure to TCDD in air. I calculate exposures by arbitrarily assigning the TCDD a concentration of 1 pg/m^3 bound to particles in the air. Shown in Table II are the geometric standard deviation, geometric mean, and expected value estimates for adult and child exposures obtained with the assigned concentration and the parameter distributions in Table I. The third column in Table II gives E*, which is the the exposure obtained using the assigned concentration and the expected values of each parameter in Table I to calculate exposure. Also listed in Table II is the 95% quantile values for each exposure estimate. One notes that for exposure outcomes, the expected value is always greater than the median. This indicates that the output distributions are skewed.

It is important to note that the geometric standard deviations for these exposure estimates are relatively large--6.2 and 6.4. We also see that the expected

Table II. Uncertainty in exposure estimates for TCDD attributable to an air concentration of 1 pg/m^3.

Pathway	GSD[a]	Geometric mean	E* [b]	Expected value	95% upper bound
		Exposure in pg/kg-d			
Meat:					
child	6.4	73	170	480	1,600
adult	6.4	48	120	320	1,000
Milk:					
child	6.2	65	170	400	1,400
adult	6.2	15	37	91	320

[a] GSD = geometric standard deviation
[b] E* is the exposure estimate that is obtained by using
  the mean value of the parameters listed in Table I.

value of exposure estimated from the Monte Carlo simulation can be significantly larger than the value E* obtained using a single estimate based on the mean value of the input parameters.

A preliminary evaluation of the sources of variance indicates that the uncertainty in exposure through the air particles-to-meat and -milk pathway is dominated by uncertainty in the biotransfer factors, B$_k$ and B$_t$, and the deposition factor.

REDUCING UNCERTAINTIES

When the magnitude of uncertainty in an exposure and risk assessment impedes the process of environmental management, decision makers must identify strategies for reducing uncertainty. Because overall uncertainty is attributable to both variability in natural systems and our ignorance about these systems, there are two obvious methods for reducing uncertainty -- improved models and expand our data. However, unless our strategy for reducing uncertainty recognizes that the cost of building new models and collecting data must be balanced by the value of the information obtained, we could squander limited resources for environmental research.

Improved Models

It seems reasonable that increasing the complexity of a model should result in better model predictions. However, it should be recognized that increasing the complexity of a model can be motivated by the need to make a model more credible. Making a model more credible does not guarantee that it is also more precise. As revealed in the example here one can increase the complexity of a model and thus improve its credibility, but without concurrently improving the precision of the model inputs one does not necessarily increase the precision of the outputs.

Figure 3 illustrates two models for predicting the deposition of a contaminant from the atmosphere to the surface of vegetation and the subsequent transfer to the fruit portion. For a risk assessment in which we want to characterize the relation between the contaminant concentration in air and the concentration in edible portion of vegetation, a two-compartment model that includes both deposition to leaf surfaces and translocation from leaves to fruits is more credible than a one-compartment model that combines all of the plant components into a single "black box." However, if our goal is to maximize precision, the two-compartment model may not be the optimum choice.

Consider the plant system shown in Figure 3 as a one-compartment system bounded by the dotted line. The rate of transfer of contaminant from air to leaf surfaces is I mg/d; the plant has a mass of M kg; and the rate constant for weathering of contaminant from the leaf surfaces is λ (1/day). For this system the steady-state concentration, C_T, of contaminant in the overall plant system is given by

$$C_T = I / (M \lambda).$$ [20]

Because we do not recognize the translocation process in this simple model, we must assume that this is also the concentration in the fruit portion of the plant.

In a two-compartment model that includes mass M_1 of the fruit compartment and rate constants k_{12} and k_{21}, which describe the rate at which a contaminant is translocated between leaf surfaces and fruit, the steady state contaminant concentration in the fruit is

$$C_1 = k_{21}I/(k_{12}M_1 \lambda).$$ [21]

In order to illustrate how lack of precision in estimating the two rate constants can affect model precision, I assign the parameters M, I, λ, and k_{12} the

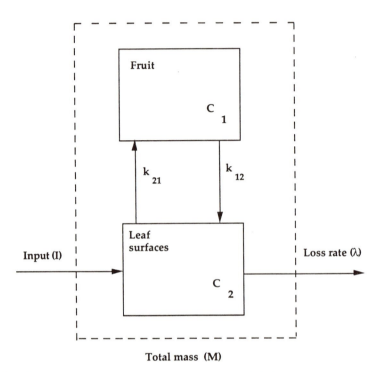

Figure 3. Illustration of a two-compartment model for representing the deposition of contaminants onto the leaf surfaces of vegetation with translocation to the fruit of that vegetation; the dotted line defines a one-compartment representation of the system.

value 1, M_1 the value 0.5, and k_{21} the value 0.25. In addition, I assume that the parameters M and M_1 have been measured with a geometric standard deviation (GSD) of 1.2, λ with a GSD of 1.5, k_{21} and I with a GSD of 2.0 and k_{12} with a GSD of 3.0. Figure 4 illustrates the spread of concentrations in fruit estimated with the simple (one-compartment) and complex (two-compartment) models. In these figures the expected value or mean of the estimated concentrations is shown by an open diamond and the geometric mean (or median) value by an asterisk. The 90 % confidence bounds are marked by an x and the 95 % confidence bounds are marked by an open circle. We can see that although the median value of fruit estimated using the "complex" model has a lower value than that estimated using the "simple" model, the mean values of fruit concentration for the two models are comparable and the complex model actually has a slightly higher value. This is attributable to the large variance in the model parameters. The complex (or more credible) model only gives a lower expected value of concentration when we increase the precision of its parameters.

More Data and Value of New Information

In addition to model improvement, data collection provides another option for reducing the uncertainty in model predictions. However, the process of collecting new information has a cost and this cost must be balanced against the value of reducing uncertainty. Evans (1985) has illustrated the use of a value-of-information framework for making a trade-off between levels of control and the need for information. Evans shows that it is possible to use the analytic framework of statistical decision analysis to determine when additional information would be beneficial and when it would not. Finkel and Evans (1987) have considered the problem of making research and control decisions using uncertain information and have proposed a model for evaluating the benefits of uncertainty reduction. They consider the total social cost of a control strategy as

$$TC = v(1-e)R \times C \qquad\qquad [22]$$

where TC is the total social cost of an environmental chemical that poses an uncertain health risk; v is the value of life, \$/death; e is the risk reduction efficiency, $0 \le e \le 1$; R is the expected health risk, death/y, in the absence of control; and C is the economic cost in \$/y of controlling the risk with efficiency e. In many situations the cost of control rises exponentially with the efficiency required. Also, it should be noted that the probability distribution of risk is likely

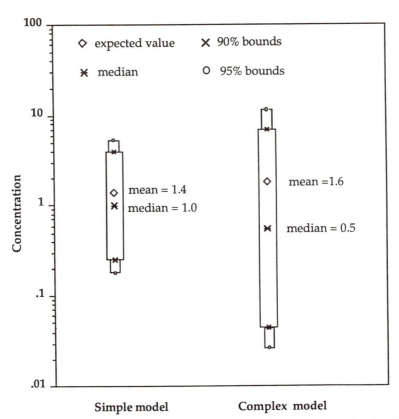

Figure 4. Uncertainty of estimated concentration of contaminant in the fruit of the plant illustrated in figure 3, where the "simple model" refers to the one-compartment representation and the "complex model" refers to the two compartment representation.

to be log normal such that the expected risk increases with uncertainty in risk as expressed in the geometric standard deviation. In situations where the expected risk is high due primarily to uncertainties in exposure, one can assign a high value to the process of reducing the expected risk by reducing uncertainty in exposure.

SUMMARY AND CONCLUSIONS

This paper describes a strategy for providing a more comprehensive assessment of the uncertainty in exposure estimates. Using TCDD as an example, I have identified pathway exposure factors (PEFs) that link ambient air concentrations to exposures through ingestion of meat and milk. A Monte Carlo simulation is applied to the data and models used to construct the PEFs. The results point out the importance of fully characterizing the uncertainty in exposure estimates. The analysis reveals that, for simple compartment models, the expected values of exposure obtained from probability distributions cannot be accurately estimated using only the mean values of the input parameters; that uncertainties in the input data limit the precision of exposure predictions to a 90% confidence range of roughly two orders of magnitude or more, and that much of the overall uncertainty in exposure is attributable to uncertainty in biotransfer factors and deposition factors. We also find that, with positive skewed input distributions, the expected value of exposure taken from the output distribution can be significantly larger than the value obtained using a single estimate based on the expected values of the input parameters. This situation can be remedied for simple models with independent inputs by recognizing that the expectation of a sum is the sum of the expectations, the expectation of a product is the product of expectations, and the expectation of quotient is the expectation of the numerator times the expectation of the inverse of the denominator. However, when we consider complex models and correlated inputs, such simple analyses become almost impossible.

The examples in this paper suggests that risk managers should be aware of the uncertainty in risk estimates and include this awareness in their decisions and their communications of risk to the public. Furthermore, the results suggest the need to focus on reducing uncertainty before we develop more sophisticated models. The PEF models presented here are simple but the precision on the output appears to be limited by uncertainty in biotransfer factors and deposition factors. Given the state of knowledge about such processes, even a highly sophisticated, multiple-compartment, time-dependent model must rely on these same parameters and will not be able to reduce this uncertainty substantially until we have a better understanding of the underlying mechanisms. This problem implies that decision makers should use an uncertainty analysis to define strategies for reducing uncertainty in risk assessment.

ACKNOWLEDGMENTS

This work was performed under the auspices of the U.S. Department of Energy (DOE) through Lawrence Livermore National Laboratory under Contract W-7405-Eng-48 with funding provided by the California Department of Health Services (CDHS), Toxic Substances Control Division through Memorandum of Understanding Agreement # 87-T0102.

REFERENCES

Bidleman, T. F. (1988), "Atmospheric Processes," *Environ. Sci. Technol.* **22**, 361-367.

Bowen, H. J. M. (1979), *Environmental Chemistry of the Elements* (Academic Press, London) pp. 49-82.

Evans, J. S. (1985), "The Value of Improved Exposure Estimates: A Decision Analytic Framework," in *Proceedings of the 78th Annual Meeting of the Air Pollution Control Association*, paper 85-33.4.

Finkel, A. M. and J. S. Evans (1987), "Evaluating the Benefits of Uncertainty Reduction in Environmental Health Risk Management," *J. Air Pollut. Control Assoc.* **37**, 1164-1171.

Hahn, G.J., and S. S. Shapiro (1967), *Statistical Models in Engineering* (John Wiley and Sons, Inc., New York , NY).

Iman, R. L., and M. J. Shortencarier (1984), *A FORTRAN 77 Program and User's Guide for the Generation of Latin Hypercube and Random Samples for Use With Computer Models*, SAND83-2365 and NUREG/CR-3624; Sandia National Laboratories, Albuquerque, NM.

International Commission on Radiological Protection (ICRP) (1975), *Report of the Task Group on Reference Man*, ICRP report 23 (Pergamon Press, New York, NY).

Jensen, D. J., and R. A. Hummel (1982), "Secretion of TCDD in Milk and Cream Following the Feeding of TCDD to Lactating Dairy Cows," *Bull. Environ. Contam. Toxicol.* **29**, 440-446.

Kenaga, E. E., (1980), "Correlation of Bioconcentration Factors of Chemicals in Aquatic and Terrestrial Organisms with Their Physical and Chemical Properties," *Environ. Sci. Technol.* **14**, 553-556.

Layton, D. W., T. E. McKone, C. W. Hall, M. A. Nelson, and Y. E. Ricker (1986), *Demilitarization of Conventional Ordnance: Priorities for Data-Base Assessments of Environmental Contaminants*, UCRL-15902, Lawrence Livermore National Laboratory, Livermore, CA.

McKone, T. E., and P. B. Ryan (1989), "Human Exposures to Chemicals through Food Chains: An Uncertainty Analysis," *Environ. Sci. Technol.* **23**, 1154-1163.

McMahon, T. A., and P. J. Denison (1979), "Empirical Atmospheric Deposition Parameters – a Survey," *Atmos. Environ.* **13**, 571-585.

Ng, Y. C. (1982), "A Review of Transfer Factors for Assessing the Dose from Radionuclides in Agricultural Products," *Nucl. Safety* **23**, 57-71.

Peterson, H. T. (1983), In *Radiological Assessment A Textbook on Environmental Dose Analysis*, Till, J. E., and Meyer H. R., Eds., Office of Nuclear Reactor Regulation, U. S. Nuclear Regulatory Commission: Washington DC, NUREG/CR-3332, Chapter 5.

Ryan, P. B., and R. Letz (©1984 to 1988), *SIMSYS Simulation System*; serial number V4.11RYAN.

Schroeder, W. H., and D. A. Lane (1980), "The Fate of Toxic Airborne Pollutants," *Environ. Sci.Technol.* **22**, 240-246.

Spengler, J. D., D. W. Docker, W. A. Turner, J. M. Wolfson, and B. G. Ferris Jr. (1981), "Long-Term Measurements of Respirable Sulfates and Particles Inside and Outside Homes," *Atmos. Environ.* **15**, 23-30.

Travis, C. C., and A. D. Arms (1988), "Bioconcentration of Organics in Beef, Milk, and Vegetation," *Environ. Sci. Technol.* **22**, 271-274.

U.S. Environmental Protection Agency (EPA) (1987), "Guidelines for Exposure Assessment," *Fed. Regist.* **51**(185), 34042.

Wallace, L. A. (1987), *The Total Exposure Assessment Methodology (TEAM) Study*, Vol. 1, EPA/600/6-87/002a, U.S. EPA, Washington D.C.

Whicker, F. W., and T. B. Kirchner (1987), "Pathway: A Dynamic Food-Chain Model to Predict Radionuclide Ingestion after Fallout Deposition," *Health Phys.* **52**, 717-737.

Wonnacott, R. J., and T. H. Wonnacott (1985), *Introductory Statistics* (John Wiley and Sons Inc., New York, NY), pp 342-350.

Yang, Y. Y., and C. B. Nelson (1986), "An Estimation of Daily Food Usage Factors for Assessing Radionuclide Intakes in the U.S. Population," *Health Phys.* **50**, 245-257.

UNCERTAINTY ANALYSIS: AN ESSENTIAL COMPONENT

OF RISK ASSESSMENT AND RISK MANAGEMENT

Rick Tyler

Health and Safety Program Specialist
California Energy Commission

Obed Odoemelam

Staff Toxicologist
California Energy Commission

Paul Shulec

Delta Environmental Consultants, Inc.

Michael Marchlik

Principal Scientist
EBASCO Services, Inc.

INTRODUCTION

Concerns over dwindling landfill capacity in California and the state's dependence on foreign energy resources have prompted several attempts to site waste-to-energy facilities in California. Most of these attempts have failed mostly as a result of public concern over potential health risks. The uncertainties associated with quantifying health risks have been a major obstacle to siting waste-to-energy facilities in California.

In recent years, quantitative health risk assessment has become the accepted method for evaluating the potential health risks associated with public exposure to toxic and/or carcinogenic pollutants. Such assessments generally consist of four steps, including:

(1) Hazard identification: The determination of which pollutants may pose a health risk and identification of what effects may result from human exposure.

(2) Dose response assessment: The determination of the relationship between the magnitude of exposure and occurrence of the health effects in question.

Municipal Waste Incineration Risk Assessment
Edited by C.C. Travis, Plenum Press, New York, 1991

(3) <u>Exposure assessment</u>: The determination of the extent of human exposure associated with a proposed project.

(4) <u>Risk characterization</u>: The description of the nature and magnitude of human health risks that could result from human exposure to pollutants emitted by the proposed project.

THE ROLE OF UNCERTAINTY ANALYSIS IN DECISION MAKING

In the context of regulatory decision making, risk assessments form the basis for weighing the health implication associated with achieving a desired goal such as disposal of solid waste. To make sound regulatory decisions, all of the available options for achieving the goal must be identified and evaluated. A risk management decision can then be made by considering the magnitude of health risks in conjunction with social benefits, environmental consequences, and the economic implications of each option. In practice, however, most risk assessments only provide analysis of the risk from the proposed project. Absence of information regarding the risk associated with alternatives, in conjunction with a large degree of uncertainty in the data used to develop risk estimates, often leaves decision makers without the information necessary to make sound risk management decisions.

Proper identification and quantification of uncertainties in risk estimates is essential to promoting public confidence in risk assessments. Such analyses allow decision makers and the public to be reasonably confident that estimates of health risks reflect the uncertainties associated with their derivation. To reduce the impact of uncertainty on the decision making process it is also essential that risk assessments compare the risk of all regulatory options using a consistent approach. This reduces the impact of major uncertainties since they are common to evaluation of all options.

Formal definitions of "risk assessment" and "risk management", and clarification of their respective roles in regulatory decision making, were first delineated by the National Research Council in 1983 (NRC 1983). In describing those roles, the NRC stated that risk assessment should be separated from risk management. More specifically, the NRC cautioned against letting political and policy decisions govern the scientific exercise of assessing risk. These recommendations later formed the basis for the formulation of federal risk assessment guidelines by the Environmental Protection Agency (Federal Register 1986).

The guidance provided by the NRC and EPA has often been misinterpreted by risk assessors to preclude them from making recommendations to decision makers regarding management of risks. Unfortunately, this misinterpretation has caused many risk assessors to focus their analysis on evaluating only the risks associated with a proposed project neglecting analysis of the risks associated with alternative approaches used to achieve the same social objectives. Thus decision makers must often base their decisions solely on uncertain risk estimates for a single option.

If instead risk assessments are used as a comparative tool for the assessment of regulatory options, debate can be focused on a choice between available options rather than acceptance or rejection of the risks associated with a specific project. It should also be

recognized that rejection of a specific proposal will often imply continuation of current methods of achieving a necessary goal.

Limiting the assessment of health risk to consideration of a single option tends to focus attention on uncertainties and analytical methods as opposed to the more appropriate determination of which regulatory option maximizes public benefit.

Most of the uncertainty associated with developing risk estimates results from uncertainties in establishing the dose response relationship and in estimation of public exposure levels. Uncertainty must therefore, be addressed in both the dose response and exposure assessment components of a risk assessment.

MAJOR SOURCES OF UNCERTAINTY ASSOCIATED WITH DOSE RESPONSE ASSESSMENTS

Current understanding of the mechanisms leading to cancer causation is very limited. It is, therefore, not presently possible to assess the cancer-causing potential of a substance directly from consideration of its chemical properties. As a result, estimation of cancer potencies for many substances must be based on results from experiments using laboratory animals as a surrogate for humans. Such experiments generally lack the sensitivity to provide statistically significant results for exposure levels which are relevant to the assessment of typical human exposures considered in risk assessments. It is therefore necessary to extrapolate from high experimental exposure levels to the much lower levels which are relevant to the usual human exposure situation. Figure 1 illustrates the extent to which extrapolation from actual experimental data is necessary to obtain information of relevance to typical human exposure situations (public exposures likely to result from normal operation of a facility).

Evaluation of dose response relationship is further complicated by the inability to experimentally determine the shape of the dose response curve for doses significantly below the exposure levels that would cause observable effects in animal experiments (NRC 1983). As a result, there are several equally plausible models for extrapolating to low doses, with different models producing risk estimates that vary by several orders of magnitude. Figure 2 illustrates the effect of using different low dose extrapolation models, on the magnitude of the unit risk obtained.

There are also many other uncertainties associated with development of dose-response relationships. For example, there are uncertainties regarding the relative susceptibilies of test animal species as compared to humans. Additional uncertainties are associated with use of high dose levels which may cause responses that would not result from lower dose levels. Such responses could result from experimental dose levels high enough to overwhelm physiological functions that may have a significant bearing on the mechanism associated with cancer causation.

The approach recommended by the Environmental Protection Agency to ensure adequate consideration of the major uncertainties associated with dose-response assessments in developing cancer potency estimates is use of data from the most sensitive test species and extrapolating to low doses using the multi-stage model (USEPA 1986). It is generally agreed that estimates of cancer potencies obtained from such an approach represent upper bound values of cancer potencies. In light of the current uncertainties about the mechanisms of cancer causation, this approach appears to be both prudent and justifiable.

While the approach described above appears justified in light of limitations on our current level of understanding of the carcinogenic process, it is likely that the significance of the uncertainties in the potency estimates will continue to be a source of controversy in the decision making process. Such controversies are likely to play an even more significant role in cases in which the environmental acceptability of projects is decided solely on the basis of the magnitude of the absolute cancer risk estimate associated with a specific project. Much of the debate regarding such uncertainties can be avoided if the risks associated with alternatives to a project are also considered using similar risk estimation methods. In many cases, the risk associated with available alternatives are of the same or greater magnitude than those associated with the proposed project.

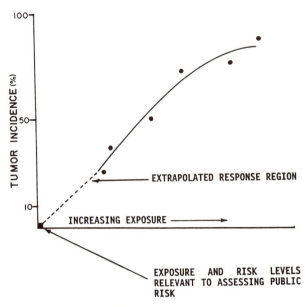

Figure. 1. Extrapolation to relevant exposure levels.

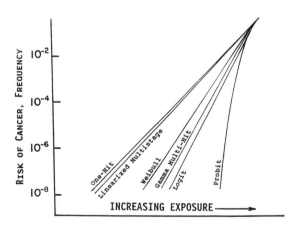

Figure 2. Differences in estimates of cancer potency using different models for low dose extrapolation. Adapted from EPA (1985).

than those associated with the proposed project.

In the case of waste management, current disposal, mainly by landfilling, is known to pose some degree of health risk to humans. In assessing the environmental acceptability of waste-to-energy projects, a risk assessor should therefore consider the risk associated with continued use of land filling for waste disposal. Additionally, the risks from energy production using fossil fuels should be compared with those from energy production by waste incineration. It is also prudent that the risk associated with other viable alternatives such as recycling be considered.

Since the use of similar risk estimation methods ensures that the uncertainties in the risks estimated for the proposed project as well as available alternatives are similar, the question of uncertainty in the risk estimation process should therefore, have diminished influence on the process of identifying the best available option.

MAJOR SOURCES OF UNCERTAINTY ASSOCIATED WITH EXPOSURE ASSESSMENTS

Over the past few years it has become recognized that many pathways of exposure must be considered in the evaluation of waste-to-energy projects in order to adequately characterize the potential health risk. In evaluating these associated health risks, the staff of the California Energy Commission evaluates potential human exposure through the following pathways:

- o   Inhalation
- o   Soil ingestion
- o   Crop ingestion
- o   Milk ingestion
- o   Livestock ingestion
- o   Surface water ingestion
- o   Fish ingestion
- o   Groundwater ingestion
- o   Maternal milk ingestion

Estimates of human exposure through each pathway are normally derived from a mathematical model that is used to simulate the transport of each emitted substance through various environmental media.

There are two major sources of uncertainty associated with this type of modeling. The first is the structural uncertainty associated with the mathematical model used to estimate exposure levels. This is the uncertainty associated with the mathematical model's ability to accurately predict actual environmental transport. Quantification of this type of uncertainty requires model validation studies. At present there is little information available to allow evaluation of this type of uncertainty.

The second source of uncertainty associated with estimating exposure is parameter uncertainty. This is the uncertainty associated with the accuracy of experimental values assigned to each of the variables or parameters used as input to the model. Uncertainty in exposure estimates can become very large as a result of parameter uncertainty. This is particularly true when estimation of exposure requires evaluation of several pathways each of which is in turn dependent on numerous input values with significant statistical variability.

To evaluate potential public exposure and also quantify the uncertainty in exposure estimates, the California Energy Commission staff relies on the "Environmental Pathways Uncertainty and Screening Model" (EPUS), developed for the Commission staff by Envirosphere Company (Envirosphere 1988). This model allows random sampling of the probability distribution for each of the input variables used in the calculation of exposure through each exposure pathway. Each set of random samples from the individual input probability distributions is input to the mathematical model and used to develop a probability distribution for the exposure estimate. Figure 3 depicts this stochastic approach.

Table 1 shows a typical pathway model used to predict exposure through the soil ingestion pathway and defines the individual input parameters for this pathway.

TABLE 1
Soil Ingestion Pathway Equation

$$DOSE = \frac{EMIS \times DOSOIL \times SIR \times GAE \times SR \times CF}{IDEPTH \times SDENS1 \times BW}$$

$$SR = \frac{1 - EXP\ (-1 \times SLAMBDA \times T)}{SLAMBDA}$$

Where: DOSE   = soil ingestion contaminant dose (mg/kg/day).
        SDENS1 = soil density (kg/m^3).
        SIR   = soil ingestion rate (kg/day).
        GAE   = gastrointestinal absorption efficiency.
        IDEPTH = soil depth for ingestion (m).
        T   = facility life (yr.)
        EMIS   = emission rate (g/sec).
        DQSOIL = atmospheric deposition factor (g/m^2/yr)/(g/s).
        SLAMBDA = contaminant half-life in soil (yr^{-1}).
        SR   = soil accumulation period (yr).
        CF   = conversion factor (converts grams to milligrams).

Source: Adapted from Nuclear Regulatory Commission (1977).

To ensure that estimates of exposure adequately reflect the related uncertainty, the Commission staff uses the upper 95 percent confidence limit of the exposure estimate for each pathway to calculate cancer risk estimates. This approach to ensuring that uncertainties are adequately reflected in risk estimates differs substantially from the approach used in development of most risk assessments. The approach currently used by most risk assessors involves the use of the plausible worst case values for each input parameter in order to maximize exposure, thus guaranteeing a health-conservative estimate.

This approach neglects the fact that most input parameters vary independently and that their variance is likely to be random in nature. This approach also implicitly requires an assumption that the variability in all input parameters is such that it results in underestimation of exposure. However, variance in some parameters will result in underestimation while variance in others will result in overestimation and many will vary only slightly from the mean resulting in only slight differences between estimated and actual risk.

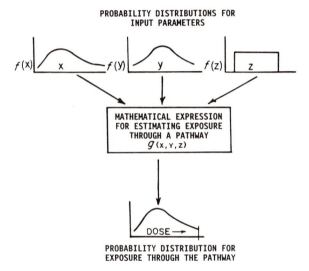

Figure 3. Parameter uncertainty analysis using a random sampling approach. Adapted from USNRC (1983).

TABLE 2

Soil Ingestion Pathway Risk Assessment
Assessment Assumptions With and Without Uncertainty Analysis

	Case 1	Case 2
	No Uncertainty Analysis Point	Uncertainty Analysis Point Estimate of
Variable	Worst Case Assumptions	Probability Distribution
BW	50	NORMAL, mean = 50, std.dev.= 6.5
SDENS1	1200	NORMAL, mean = 1500, std.dev. = 97.1
SIR	0.2	TRIANGULAR, endpoints = 0, 0.2 midpoint = 0.1
GAE	1.0	TRIANGULAR, endpoints = 0, 1.0 midpoint = 0.5
IDEPTH	0.01	UNIFORM, endpoints = 0.01, 0.05
T	70	HISTOGRAM, 25 - 40, 50%;) 40 - 50, 40%; 50 - 70, 10%
EMIS	1.05E-02	LOGNORMAL, median = 9.25E-04 geo. std. dev. = 2.2
DQSOIL	0.092	LOGNORMAL, median = 0.0054 geo. std. dev. = 2.5
SLAMBDA	0.00061	LOGNORMAL, median = 0.006 geo. std. dev. = 2.1
DOSE	2.21E-05	Median 2.54E-09 95% tile 2.35E-08 Max 1.08E-07 Min 9.92E-12

There are significant differences in the magnitude of the exposure estimates obtained using each of these two methods. Table 2 shows the results of comparing the two approaches in evaluation of arsenic exposure through the soil ingestion pathway. These results indicate a difference of approximately 3 orders of magnitude in the exposure estimates obtained using these two methods. Larger differences should be expected in exposure estimates involving more complex pathway models and for pollutants with greater uncertainty in environmental transport parameters. This example shows that consistent use of conservative assumptions for every input to an exposure pathway model can result in significant overestimation of risk.

Differences of this magnitude which result solely from application of different analytical approaches applied to the same experimental data, can also seriously undermine the confidence of a decision maker who must rely on such analyses.

CONCLUSIONS

Since uncertainty analysis will continue to play a significant role in promoting public acceptance of the results of risk assessments, it is essential that such analyses provide an accurate characterization of risk associated with human exposure to carcinogens. Consistent use of worst case assumption to address uncertainty in estimates of exposure can result in significant overestimation of actual risk. Such overestimation of risk can be avoided by using a stochastic approach in risk analysis.

The use of risk assessment to compare regulatory options as opposed to its use in characterizing the absolute risks from a single regulatory option would, 1) limit the potential debate over the uncertainties associated with risk estimates, and 2) help focus the decision making process on the choice of the best option for minimizing the potential public health risks.

REFERENCES

Envirosphere Company. 1988. Environmental pathways uncertainty and screening model user guide. Prepared for the California Energy Commission, June, 1988.

Environmental Protection Agency (EPA). 1986. Guidelines for carcinogen risk assessment. Federal Register Vol. 51, pp. 33992-34003.

Environmental Protection Agency (EPA). 1985. Toxicology Handbook. Principles related to hazardous site investigations. p. 6-11. EPA TR 693-21A.

National research Council (NRC). 1983. Risk assessment in the federal government. Managing the process. National Academy Press. Washington, D.C.

Nuclear Regulatory Commission (USNRC). 1977. Regulatory Guide 1.109. U.S. Nuclear Regulatory Commission, Washington D.C.

Nuclear Regulatory Commission (USNRC). 1983. Radiological Assessment. U.S. Nuclear Regulatory Commission Washington, D.C., NUREG/CR3332.

AMBIENT AIR CONCENTRATIONS OF POLYCHLORINATED DIBENZO-P-DIOXINS AND DIBENZOFURANS IN OHIO: SOURCES AND HEALTH RISK ASSESSMENT [Ω]

Sylvia A. Edgerton [§], Jean M. Czuczwa [Φ], and Jerry D. Rench

Battelle Columbus Division, 505 King Avenue Columbus, Ohio 43201

Robert F. Hodanbosi and Paul J. Koval

Ohio Environmental Protection Agency, 1800 Watermark Drive, Columbus, Ohio 43215

INTRODUCTION

Concentrations in the environment of polychlorinated dibenzo-p-dioxins (PCDD), and related compounds such as polychlorinated dibenzofurans (PCDF), have recently been the cause of great concern due to their suspected high toxicity. General low background concentrations of PCDD/PCDF are found in the atmosphere, sediments and the human population[1]. Evidence in the sediment record suggest that there have been increases in PCDD/PCDF pollution since 1940[2]. In this study, we present the results of a program to determine the sources, occurrence and effects of PCDD and PCDF concentrations in ambient air in Ohio at several locations.

Animal studies have demonstrated that 2,3,7,8-tetrachlorodibenzo-p-dioxin (2,3,7,8-TCDD) is teratogenic and fetotoxic, and along with 1,2,3,6,7,8 and 1,2,3,7,8,9-hexachlorodibenzo-p-dioxin (HxCDD) is a probable human carcinogen[3]. However, epidemiologic studies of 2,3,7,8 TCDD exposure from the defoliant Agent Orange, such as those

---

[Ω] Reprinted with permission from **Chemosphere,** Vol. 18, Nos. 9/10, pp 1713-1730, Copyright 1989, Pergamon Press, plc.

[§] Currently located at the Environment and Policy Institute, East-West Center, Honolulu, HI 96848

[Φ] Currently located at Babcock & Wilcox, 1562 Beeson St., Alliance, Ohio 44601

conducted on Vietnam veterans, have produced conflicting
results. Existing data have shown the prevalence of PCDDs and
PCDFs in human adipose tissue in the United States, Canada
Sweden, and Japan. The existence of these compounds in air, in
drinking water, and in sediments remains a cause for public
concern. 2,3,7,8-TCDD is a compound with no known major
natural sources, but is formed as a by-product in the
manufacturing of chlorinated phenolic compounds and as a
product of many combustion processes[4]. It is generally
resistant to degradation once adsorbed onto soil, with a half
life of 10-12 years, and bioaccumulates in fish and mammals[5].
Studies by Czuczwa and Hites[2,6] have confirmed the importance
of combustion as a source of PCDD and PCDF in sediments from
Lake Erie and Siskiwit Lake (Isle Royale).

REVIEW OF PCDD/PCDF MEASUREMENTS IN AMBIENT AIR

A large body of literature has developed containing
information on PCDD and PCDF content in municipal incinerator
ash and stack gases. Limited measurements exist to date on
PCDD/PCDF concentrations in ambient air. This is primarily due
to the complex analytical requirements necessary to achieve low
detection limits and avoid interferences from other chlorinated
compounds. A literature search was conducted to review all
ambient air measurements to date. Table 1 shows a comparison
of PCDD/PCDF congener concentrations measured in ambient air at
several sites[7-11].

In Table 1, one notes some trends regarding PCDD/PCDF
concentrations in ambient air: (1) the highest levels are
detected in urban and industrial areas; (2)   urban levels are
approximately 10 times suburban levels; (3) 2,3,7,8-TCDD is
generally low (<100 fg/m^3, except for areas where high levels
of documented dioxin contamination are present (e.g. Times
Beach, MO); (4) 2,3,7,8-TCF is usually higher than the
corresponding 2,3,7,8-TCDD; and (5) high levels detected in an
automobile tunnel[1] suggest automobile inputs to the levels of
PCDD/PCDF in air.

TABLE I. LEVELS OF PCDF IN AIR (fg/m^3)

Location	Ref.	Type	2378 TETRA -CDF	TOTAL TETRA -CDF	TOTAL PENTA -CDF	TOTAL HEXA -CDF	TOTAL HEPTA -CDF	OCTA -CDF	TOTAL PCDD/ PCDF
Bloomington, IN[a]	7	suburban	--[d]	340	200	120	93	36	2200
Trout Lake, WI	7	rural	--[d]	83	67	31	12	6.0	510
W. Germany (W)	8	industrial	90	1600	1600	790	1000	490	8700
W. Germany (W)	9	industrial	370	7600	5600	2700	2500	780	37000
W. Germany (W)	9	suburban	40	360	510	180	100	N.D.	2900
W. Germany (W)	9	tunnel	450	4900	3900	1600	1600	N.D.	29000
Netherlands	10	1 km/MWI	--[c]	5000	4800	5200	--[c]	4800	47000
Netherlands	10	2 km/MWI	--[c]	400	600	2100	--[c]	400	9100
Missouri	11	Superfund site	--[c]	--[c]	--[c]	--[c]	--[c]	--[c]	--[c]

EXPERIMENTAL METHODS

During November and December of 1987 six air samples were collected in Ohio for determination of PCDD/PCDF. The locations of the sampling sites relative to potential sources are shown in Figure 1. Two consecutive three day samples were collected at an industrial site, 3/4 km from a municipal refuse derived fuel (RDF) power plant and <1/4 km from a sewage sludge incinerator in Columbus, Ohio (COL1, COL2). Two one week samples were collected coincidently in Akron, Ohio (AK1, AK2), 2 km downwind of the municipal incinerator at an urban site. A three day sample was collected at a high traffic density highway site in central Columbus (17 and I71). A one week sample was collected at a rural background Ohio site, Waldo, which is approximately 45 km north of Columbus.

Samples were collected with a medium volume air sampler (rate of ~ 0.2 m^3/min) fitted with a quartz fiber filter (Pallflex 2500 QAT-UP) and polyurethane foam (PUF) plug. A polyurethane foam plug (PUF) and filter for each sample (PUF and filter together) were placed into Soxhlet extraction thimbles and spiked with known concentrations of nine $^{13}C_{12}$ labelled specific PCDD/PCDF isomers for quantitative analysis. The air samples were extracted in a Soxhlet apparatus with benzene for 18 hours. Also processed with the samples were four QA/QC samples, including a laboratory blank sample to demonstrate freedom from contamination; a laboratory native spike sample (spiked with known amounts of specific PCDD/PCDF isomers) to determine the accuracy of the measurements; and two field blank samples.

The benzene extracts were concentrated and processed on multilayered silica gel columns containing silica gel, sulfuric acid on silica gel, and sodium hydroxide on silica gel. The resulting solutions were concentrated, solvent exchanged, and processed on activated basic alumina columns. The eluates were collected, concentrated to near dryness, and an absolute recovery standard, 2,3,7,8-TCDD-$^{37}Cl_4$, was added. All

TABLE I. CONTINUED

Location	Ref.	Type	2378 TETRA -CDD	TOTAL TETRA -CDD	TOTAL PENTA -CDD	TOTAL HEXA -CDD	TOTAL HEPTA -CDD	OCTA -CDD	TOTAL PCDD/ PCDF
Bloomington, IN[a]	7	suburban	--[d]	5.3	51	140	360	890	2200
Trout Lake, WI	7	rural	--[d]	0.3	6.1	52	93	160	510
W. Germany (W)	8	industrial	N.D.	N.D.	510	610	1140	980	8700
W. Germany (W)	9	industrial	100	2100	1100	4100	4800	6100	37000
W. Germany (W)	9	suburban	20	100	50	740	440	370	2900
W. Germany (W)	9	tunnel	60	230	1900	5300	3400	6400	29000
Netherlands	10	1 km/MWI	--[c]	1500	3400	4000	5200	13300	47000
Netherlands	10	2 km/MWI	--[c]	100	500	2000	2100	900	9100
Missouri	11	Superfund site	1000	--[c]	--[c]	--[c]	--[c]	--[c]	--[c]

a Average of 55 air samples.
c Individual congener not reported.
d Individual congener not measured.

Figure 1. Map of Ohio showing sampling sites and potential PCDD/PCDF sources, municipal solid waste incinerators (MSW), sewage sludge incinerators (SSI), and hazardous waste incinerators (HWI).

+ Sampling Sites
• MSW
■ SSI
□ HWI

solutions were stored at 0°C and protected from light until analyzed.

The extracts were analyzed and quantified for PCDD/PCDF by combined capillary column gas chromatography/high resolution mass spectrometry (HRGC/HRMS). The HRGC/HRMS system consists of a Carlo Erba Model 4160 gas chromatograph interfaced directly into the ion source of a VG Model 7070 high resolution mass spectrometer. The mass spectrometer was operated in the electron impact (EI) ionization mode at a mass resolution of 9,000-12,000 (M/※M, 10% valley definition). All HRGC/HRMS data were acquired by multiple-ion-detection with a VG Model 11-250 Data System.

RESULTS

The results from the PCDD/PCDF analyses are shown in Table II. The results are given in femtograms (10^{-15} grams) per cubic meter (fg/m^3). These data are the actual measured levels in the samples and have not been corrected for laboratory or field blank levels (both laboratory and field blanks showed no detectable amounts of PCDD/PCDF). A detection limit is listed, in parentheses, for samples in which a particular chlorine congener class was not detected. The detection limit was calculated by comparison of the peak height of the internal standard to 2.5 times the peak height of the PCDD/PCDF signal. The achievable detection limits in environmental samples will be shown later in this paper to play a key role in estimating health risk to PCDD/PCDF exposures.

In general, hepta-CDD, and octa-CDD/CDF were the most abundant PCDD and PCDF in the air samples. The Akron AK1 extract was analyzed twice by HRMS to measure the reproducibility of the HRMS determination. The reproducibility was approximate ± 20% except for very low level isomers (<100 fg/m^3). The samples Akron AK1 and Akron AK2 were sampled at the same time, and thus indicate reproducibility of overall sampling. Reproducibility for the two Akron samples was approximately 20-30%, which is typical for environmental samples.

No 2,3,7,8-TCDD were detected in the Ohio air samples, with average detection limits of less than 240 fg/m^3. The recoveries of the internal standards averaged 100 +/- 11%. These recoveries indicate excellent control of the analytical procedures. Four of the polyurethane foam plugs were spiked with 10 ng of 1,2,3,4-tetra-CDD-$^{13}C_{12}$ before sample collection to determine the collection efficiency of the samplers. The recoveries of the 1,2,3,4-tetra-CDD-$^{13}C_{12}$ averaged 84 +/- 4.9%, suggesting only minor losses of the lower chlorinated (more volatile) PCDD/PCDF.

DISCUSSION

PCDD/PCDF Levels in Ohio Air Samples

The PCDD/PCDF levels found in the Ohio air samples are similar to ambient air levels reported previously at other locations. No 2,3,7,8-TCDD was detected in any of the air samples. The total PCDD/PCDF ranged from 1900-9900 fg/m^3. While the total PCDD/PCDF concentrations measured in this study are somewhat higher than the 450 fg/m^3 measured at a remote rural site in Wisconsin (see Table 1), the levels at the 17th Street and 1-71 Columbus Site, and at the rural Waldo site are similar to those measured in Bloomington, Indiana and at a rural site in Norway. The ambient concentrations of total PCDD/PCDF in samples collected near municipal incinerators (Akron and Columbus) are about twice that found at the background sites, but average only about 20 percent of the total PCDD/PCDF found at urban and industrial sites in Europe.

Analysis of Sources

A literature search was conducted to develop congener profiles representative of each source type. A large body of data exists demonstrating the tremendous variability if PCDD and PCDF emissions from municipal incinerators[12]. We used a pattern recognition program and classification scheme to identify incinerators with similar congener profiles in emission measurements, and developed a "generic" municipal incinerator profile for those incinerators which fall within this group. We began with the worldwide database compiled by Roy F. Weston, Inc. for municipal solid waste incinerator (MSWI) PCDD/PCDF emissions[13]. Total PCDD in emissions from 32 incinerators varied from <1 to 33,047 ng/dry standard cubic meter (ng/DSCM). In order to develop emission factors from this database, the authors narrowed the database by eliminating incinerators (1) without heat recovery, (2) experiencing abnormal conditions during testing, (3) using refuse derived fuel (RDF), and (4) characterized as smaller modular facilities. The reduced database included 13 incinerators and the range of total PCDD in emissions was narrowed to 1.3 to 185 ng/DSCM. This reduced database was used to construct the congener profile for the municipal incinerator source in this study. Because the municipal incinerators of interest in Ohio burned refuse derived fuel (RDF), three RDF incinerators for which emissions data were available were added to the database. Four of the incinerators in the Weston database were eliminated based on incomplete congener profile emission data. A list of the remaining MSWIs are shown in Table III, along with the total PCDD reported in their emissions.

The similarity of congener profiles among the incinerators listed in Table III was examined by conducting a principal components analysis (PCA) and classification, using SIMCA-3B[14] software, on the normalized congener concentration

TABLE II. LEVELS OF PCDF IN OHIO AIR SAMPLES (fg/m³) [a]

Sample Number	2378 TETRA -CDF	TOTAL TETRA -CDF	12378 PENTA -CDF	23478 PENTA -CDF	TOTAL PENTA -CDF	123478 HEXA -CDF	123678 HEXA -CDF	234678 HEXA -CDF	123789 HEXA -CDF	TOTAL HEXA -CDF	1234678 HEPTA -CDF	1234789 HEPTA -CDF	TOTAL HEPTA -CDF	OCTA -CDF
AK1	200	990	26	32	530	100	55	(36)	39	600	250	(35)	390	190
AK1 rep	200	1200	33	42	660	53	48	(21)	36	700	240	(22)	390	170
AK2	190	1500	29	34	580	95	92	(5.3)	20	560	220	31	370	180
COL1	320	1900	32	(23)	690	60	92	(28)	38	370	200	(15)	260	(310)
COL2	490	3800	57	89	1300	270	190	(12)	120	1200	470	(28)	640	210
17 & I-71	(130)	(130)	(36)	(36)	(36)	(34)	(34)	(34)	(34)	100	87	(13)	150	(160)
Waldo	130	890	21	(33)	500	98	14	(8.3)	97	510	220	19	290	77

a A detection limit is listed in parentheses for samples in which a particular chlorine congener class was not detected.

Sample Number	2378 TETRA-CDD	TOTAL TETRA-CDD	12378 PENTA-CDD	TOTAL PENTA-CDD	123678 HEXA-CDD	123478 HEXA-CDD	123789 HEXA-CDD	TOTAL HEXA-CDD	1234678 HEPTA-CDD	TOTAL HEPTA-CDD	OCTA-CDD	TOTAL PCDD/PCDF
AK1	(200)	(200)	(270)	(270)	52	35	50	600	520	1000	1000	5300
AK1 rep	(160)	(160)	(110)	(110)	53	55	26	630	530	1100	1200	6000
AK2	(12)	180	(34)	100	53	32	17	630	570	1100	1200	6400
COL1	(820)	(820)	(60)	(60)	(28)	(28)	(28)	430	260	410	510	4600
COL2	(240)	(240)	(47)	(47)	78	(39)	64	780	520	1000	1100	9900
17 & I-71	(150)	(150)	(82)	(82)	(32)	(32)	(32)	150	320	560	960	1900
Waldo	(58)	(58)	(33)	(33)	25	31	25	330	240	480	500	3600

a A detection limit is listed in parentheses for samples in which a particular chlorine congener class was not detected.

TABLE III. MUNICIPAL INCINERATOR EMISSIONS DATA[a]. THE SEVEN INCINERATORS
BETWEEN THE TWO LINES WERE USED TO CONSTRUCT THE MUNICIPAL WASTE
INCINERATOR CONGENER PROFILE USED IN THE CMB MODEL.

INCINERATOR	PCA OBJECT #	TOTAL PCDD (NG/DSCM)[b]
Marion County (USA)	7	1.3
Hogdalen (Sweden)	6	6.5
Tulsa County (USA)	4	22
Westchester (USA)	9	24
Wurzburg (Germany)	3	25
Pittsfield (USA)	1	36
North Andover (USA)	8	122
Zurich (Switzerland)	2	171
Saugus (USA)	5	182
Albany (USA)	10	305
Niagara (USA)	23	853
Hamilton (Canada)	34	4,259

a Roy F. Weston, Inc. Database[13]   b Dry standard cubic meter

data.   The   PCA decomposes the data matrix X as:

$$X = 1 \cdot \underline{x} + T \cdot P + E \qquad (1)$$

where the n x F score matrix T describes the projection of the
n object points on the F dimensional hyperplane defined by the
F x p loading matrix P.   Further details of the SIMCA-3B
calculations can be found in Wold and Sjostrom[15].   By initially
fitting the data to two principal components and plotting the
individual factor scores of each object (incinerator) for each
component, a two-dimensional "window" of the data is
constructed.   Figure 2 shows the results of a PCA score plot
for the 12 incinerators listed in Table III.   Incinerators 6,
7, 10, 23 and 34 appear to be outliers.   The Hamilton RDF
incinerator (object 34) was operating under unsteady burning
conditions and subsequently reported very high PCDD/PCDF
emissions.   The Niagara Falls RDF incinerator (object 23) was
operating under unsteady burning conditions and subsequently
reported very high PCDD/PCDF emissions.   The Niagara Falls RDF
incinerator (object 23) exceeded the New York State particulate
emissions limit during the tests.   The Albany RDF incinerator
(object 10) contained high total PCDD, indicating potentially
abnormal operations.   A description of the test conditions[16]
revealed abnormal operation of the electrostatic precipitators
during the test period.   The individual concentrations measured
at the Marion incinerator (object 7) were so low, and so
variable, the validity of the data may be questionable where
the measured concentrations approach the detection limits.   The
Hogdalen incinerator (object 6) contained the next lowest PCDD
concentrations after Marion in the Weston database, and could
conceivably have approached detection limits for some
congeners, although specific data on the Hogdalen tests were
not available.

    Incinerators (objects 6, 7, 10, 23 and 34 were deleted
from the model and new principal components were calculated to
model the other 7 incinerators as a "class".   The SIMCA
classification program was then used to relate each object

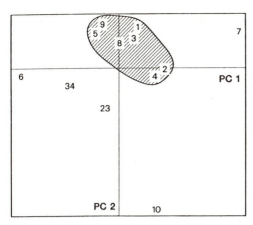

Figure 2. Principal components score plot for municipal waste incinerator PCDD/PCDF congener data, comonent 1 plotted against comonent 2. The object numbers correspond to the incinerators listed in TABLE II. Incinerators 1-5, 8, and 9 form a similarity cluster.

(incinerator) data vector (xj) to the class model and calculate a residual standard deviation (RSD) for each object. The distance from the object point to the class model is calculated as:

$$x_j - x_g = t_j P_g + e_j \qquad (2)$$

where the class g parameters P_g are fixed and the t_j parameters are calculated by linear regression. The standard deviation of the residuals, e_j, corresponds to the distance between the object and the class. Each object can be tested to see if it belongs to class g by calculating a tolerance interval around the class. This interval is based on a modified F-test based on (M-A) and (N-A-1)(M-A) degrees of freedom, where M is the number of variables, N is the number of objects in the class, and A is the number of components used to model the class[15]. Using this test, incinerators 6, 7, 10, 23 and 34 do not belong to the modeled class, which includes the other 7 incinerators were then averaged and used as the MSW incinerator profile in the source apportionment.

Profiles for three kraft paper mills, three sewage sludge incinerators and one hazardous waste incinerator were obtained from the EPA National Dioxin Study Tier 4 database[12]. Data for an additional sewage sludge incinerator profiles in the EPA data base were omitted due to significantly higher total PCDD/PCDF emissions, by a factor of 10, which indicated potentially abnormal operations. The other three sewage sludge incinerator profiles, and the three kraft paper mill profiles were normalized and averaged. The resultant profiles were used in the source apportionment. While there has recently been some indication of automobiles as a source of PCDD/PCDF, no quantitative profiles have yet been published. One of the ambient monitoring sites was selected in an area of high traffic density (highway with >80,000 cars/day) and known elevated lead concentrations. It was hoped that data from this site could be used to indicate the contribution of automobiles and other mobile sources to PCDD/PCDF concentrations in ambient urban air.

A PCA analysis was conducted for the 3 kraft paper mill samples, the 3 sewage sludge incinerator samples and the 7 municipal incinerator samples selected to represent each class. The resulting principal components score plot is shown in Figure 3. The object classes can be easily separated visually, indicating distinct congener profiles typical of each source.

A principal components analysis was conducted on the normalized source objects and included the normalized ambient air data, both from Ohio and from other locations as additional objects. Two significant principal components were found (using the CSVAL validation routine) and the individual factor scores for each component were plotted on a two dimensional score plot. Figure 4 shows the factor score plot, where the kraft paper mill objects have been left out because this source was later found not to be a contributing one. The variables used to determine the components were 2378-TCDF, tetra-CDF, penta-CDF, hexa-CDF, hepta-CDF, hexa-CDF, octa-CDF, hexa-CDD, hepta-CDD and other octa-CDD. Because the 2378 TCDD, other TCDD, and penta-CDD concentrations were always below the detection limit in the ambient samples, these variables were not used. The variables with the most modeling power were octa-CDD, tetra-CDF and penta-CDF. The other variables played less of a role in separating the source clusters.

The two Columbus samples (COL1 and COL2), collected 3/4 km from the municipal incinerator and <1/4 km from the sewage

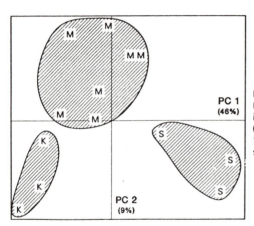

Figure 3. Principal components score plot for PCDD/PCDF congener data for municipal waste incinerators (M), sewage sludge incinerators (S) and kraft paper mill boilers (K), component 1 plotted against component 2. The three source types each form a similarity cluster.

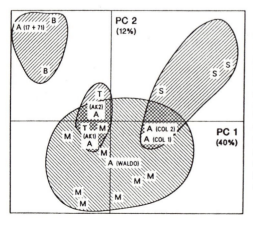

Figure 4. Principal components score plot for ambient and source samples, component 1 plotted against component 2 for: M, municipal waste incinerator source data; S, sewage sludge incinerator source data; T, ambient data from a traffic tunnel (ref. 9); B, background ambient data for Bloomington, IN and Trout Lake, WI (ref. 7),; and A, ambient data for Ohio. Columbus COL1 and COL2 show similarity with the M and S objects, Akron AK1and AK2 show similarity with the M and T objects, and urban background (17 and 71) shows similarity with the B objects. The Waldo sample is in the M cluster.

sludge incinerator, are included in both source clusters. The
two Akron samples (AK1 and AK2), collected 2 km downwind of the
municipal incinerator, are also found in the municipal
incinerator cluster and are close to ambient air samples
collected in tunnel studies[1],[18], suggestive of some automotive
influence. The sample collected at the high traffic density
site (highway traffic) in Columbus (17th St. and I-71)
contained the lowest concentrations of PCDD/PCDF, is similar to
background samples collected at Trout Lake, WI and Bloomington,
IN[7], and lies somewhat near the tunnel samples on the plot.
The winds were predominately from the northwest, with no
incineration sources upwind, during the collection period
(2 days) for this sample. The Waldo site is a rural one, and
the PCDD/PCDF results of this sample are surprisingly high
relative to the other sites, since there are no nearby sources.
The profile for the Waldo sample is almost identical to the
profile constructed for municipal incinerators (correlation
r = .94) as shown in Figure 5. The winds were variable over
the sampling period and long distance contributions from many
parts of Ohio may have contributed to this sample. This sample
may represent a regional background concentrations with
contribution from all Ohio incinerators.

     The congener profiles developed above were used in a
chemical mass balance model (CMB 5.9)[19] to apportion the
sources of PCDD/PCDF at ambient sites in Ohio. The results of
the CMB model applications are shown in Table IV. Kraft paper
mill boiler, and industrial and hazardous waste boiler profiles
did not fit into the CMB model for any of the samples. The CMB
results agree well with the placement of the ambient samples in
the source clusters of the PCA score plots (Figure 4). The
statistical parameters R^2 and C^2, the T-Stat values for the
sources, the ratios of the predicted to measured mass of each
compound, and the ratio of predicted to measured sum of all
compounds indicate that the model predicts the sources well.
The location of the sampling sites relative to nearby sources
and the correlation with prevailing winds during the sample
periods also confirm the likelihood that the predicted source
types are correct.

     The results from this study should considered qualitative
since actual profiles from Ohio sources were not used to fit
the model. We have shown that pattern recognition techniques
are very useful for identifying similar samples and
constructing source profiles for similar source types. The
consistency of the results between the PCA and the CMB models
applied to the Ohio air samples are encouraging and suggest
that the apportionment of the sources of PCDD and PCDF in the
atmosphere using congener profiles is feasible and can help to
identify the major sources of these compounds to the
atmosphere.

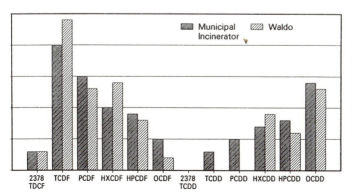

Figure 5. The histogram
demonstrates the
similarity in congener
profile for the Waldo
sample and the profile
developed for the
municipal waste
incineration source.

261

TABLE IV. CHEMICAL MASS BALANCE MODEL RESULTS: SOURCE
CONTRIBUTIONS TO PCDD/PCDF IN OHIO

LOCATION	SOURCE CONTRIBUTION		R^2	x^2	% OF MASS EXPLAINED
AK1	Municipal Incinerator Urban Background	72 ± 6 % 28 ± 11 %	0.95	1.6	111
AK2	Municipal Incinerator Urban Background	74 ± 6 % 26 ± 7 %	0.93	2.5	100
COL1	Municipal Incinerator Sewage Sludge Incin.	71 ± 9 % 29 ± 7 %	0.93	3.0	103
COL2	Municipal Incinerator Sewage Sludge Incin.	81 ± 9 % 19 ± 7 %	0.93	3.1	100
WALDO	Municipal Incinerator	100 ± 6 %	0.95	1.7	100

Dispersion Modeling

A dispersion model was used to compare the measured
concentrations with modeled concentrations using average
emission estimates from the literature for the suspected
sources in the Columbus sample---the RDF incinerator and the
sewage sludge incinerator. The emissions were converted to
toxic equivalents (TE) using the method of Bellin and Barnes[20],
which converts the PCDD/PCDF isomer concentrations to a total
toxicity equivalent to 2,3,7,8-TCDD. The emission factor used
for the municipal incinerators was 5×10^{-6} g/sec toxic
equivalents[21]. The factor used for the sewage sludge
incinerator was 6.3×10^{-8} g/sec toxic equivalents[22,23].

The ISC-ST (EPA UNAMAP-6)[24] short term dispersion model
was applied using meterological data available during the COL2
sampling period. The predicted TE at the receptor was 91 fg/m^3
during the sample period, while the measured TE was 133 fg/m^3.
This relatively close agreement between the modeled and
measured concentration at the receptor provided confidence in
the emission estimates. These emission factors were then used
in a long-term dispersions model to predict annual average
concentrations.

The ISC-LT long term dispersion model was used to estimate
annual average TE concentrations. Four years of combined
meteorological data from the Columbus area (Star data from the
National Weather Service) was used as input to the model. The
sewage sludge incinerator maximum impacts were close to the
source (less than 1 km) due to its relatively low stack height
(63 ft), whereas the municipal power plant maximum impacts were
further away (about 5 km) due to the high stacks (272 ft).
Figure 6 shows isopleths of the predicted annual average TE
concentrations in the Columbus area.

Health Risk Assessment

Potential health risks from inhalation of ambient air in
Ohio were calculated based on the toxic equivalency method
Bellin and Barnes[20]. This method assumes that a reasonable
estimate of toxic risks associated with mixtures of PCDDs and
PCDFs can be made by taking into account the relative toxicity

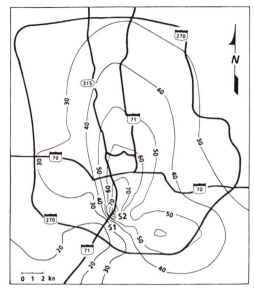

Figure 6. Map of Columbus, Ohio showing major roads and isopleths of modeled annual average concentration of PCDD/PCDF in toxic equivalents (fg/m3). S1 and S2 give the location of the municipal and sewage sludge incinerators, respectively.

of these compounds.   The concentrations of each homologue and/or congener group present in the mixture is multiplied by a toxicity equivalence factor (TEF) associated with that group. At one extreme, all homologues could be considered as toxic as 2,3,7,8-TCDD, which has a TEF of one.

An interesting problem associated with the calculation of health risks from PCDD/PCDF concentrations in the ambient air is associated with the interpretation of detection limits.   The most conservative approach assumes that if a homologue is not detected, the concentration could be as high as that of the detection limit.   This assumption is not warranted in the case of 2,3,7,8-TCDD, for example, where concentrations in the atmosphere are expected to be considerably lower than the achievable detection limits.   Because the health risks are heavily weighted toward the concentration of 2,3,7,8-TCDD, and if the concentration is assumed equal to the detection limits or even one half the detection limits, the risk becomes dependent solely upon the attainable detection limits of the analytical method and the signal/noise ratio obtained for each individual sample.   For example, the detection limits shown for 2,3,7,8-TCDD in Table II are 200 fg/m$_3$ for sample AK1 and 12 fg/m^3 for sample AK2.   Because these samples were collected coincidentally, it is unreasonable to believe that one would contain 16 times the levels of 2,3,7,8-TCDD, especially when the samples have similar concentrations of the other homologues.   Still, these two samples would yield very different calculated risks.

A preferred method for calculation of risks for low level samples would be the estimation of the probable concentrations of 2,3,7,8-TCDD based on the concentrations of the other congener/or homologue groups and based on knowledge of the potential contributing sources or background environmental samples.   Table V shows the toxic equivalent calculations for the ambient samples collected in Ohio based on the assumptions: (A) of zero concentration where the homologue is below the detection limits; (B) that concentrations are one half the detection limits; (C) that the concentrations are equal to the detection limits; (D) that the concentrations are equal to the average detection limit for that homologue during the study;

detection limits; (C) that the concentrations are equal to the detection limits; (D) that the concentrations are equal to the average detection limit for that homologue during the study; and (E) that the 2,3,7,8-TCDD/2,3,7,8-TCDF ratio in ambient air is proportional to that found in the estimated source emissions.

The estimated health risk is calculated from these toxic equivalents based on the cancer unit risk factor (ql*) obtained in the Dow Chemical study[25] and published by the U.S. EPA[3] and using average daily inhalation rates. The value for ql* for inhalation for an adult male is 3.3×10^{-8} $(fg/m^3)^{-1}$. This factor is multiplied by the concentration to give the risk factor. Risk factors for the Ohio samples are shown in Table VI using the assumptions for detection limits as stated for Table V.

TABLE V.   TCDD EQUIVALENT EXPOSURE LEVELS FOR SELECTED SITES IN OHIO
(See text for definitions of A-E)

DIOXIN EQUIVALENTS	A	B	C	D	E
AK1	37.4	205.1	272.7	372.8	115.2
AK2	36.7	51.0	269.5	64.4	54.0
COL1	40.0	467.9	278.9	895.9	83.0
COL2	81.4	213.9	314.9	346.5	133.1
COLS 17th & I-71	0.4	96.4	255.4	215.7	10.9
WALDO	25.7	63.0	258.0	100.3	83.3

TABLE VI. TCDD CANCER RISK FOR SELECTED SITES IN OHIO
(See text for definitions of A-E)

DIOXIN RISK	A	B	C	D	E
AK1	1.2×10^{-6}	6.8×10^{-6}	9.0×10^{-6}	1.2×10^{-5}	3.8×10^{-6}
AK2	1.3×10^{-6}	1.7×10^{-6}	8.9×10^{-6}	2.1×10^{-6}	1.8×10^{-6}
COL1	2.7×10^{-6}	1.5×10^{-5}	9.2×10^{-6}	3.0×10^{-5}	2.7×10^{-6}
COL2	1.4×10^{-6}	7.1×10^{-6}	1.0×10^{-5}	1.1×10^{-5}	4.4×10^{-6}
COLS 17th & I-71	8.5×10^{-8}	3.2×10^{-6}	8.4×10^{-6}	7.1×10^{-6}	3.6×10^{-7}
WALDO	8.5×10^{-7}	2.1×10^{-6}	8.5×10^{-6}	3.3×10^{-6}	2.8×10^{-6}

TABLE VII. CUMULATIVE POPULATION EXPOSURE TO PCDD/PCDF
IN AMBIENT AIR IN COLUMBUS, OHIO CALCULATED FROM
ANNUAL AVERAGED MODELED CONCENTRATIONS

Toxic Equivalents (TE)	Population (thousands) Exposed to TE $\geq$ Given Value	Current Risk Factor	Proposed Risk Factor
70	12	2.3×10^{-6}	1.4×10^{-7}
60	16	2.0×10^{-6}	1.2×10^{-7}
50	112	1.6×10^{-6}	9.6×10^{-8}
40	221	1.3×10^{-6}	7.8×10^{-8}
30	447	9.9×10^{-7}	5.9×10^{-9}

Risk calculations for annual average ambient exposures in the Columbus area as determined from a population density mapping onto the TE isopleth map shown in Figure 6 are given in Table VII. The U.S. EPA has proposed a new risk factor[26] for 2,3,7,8-TCDD, and the health risks using this factor are shown also. A health risk of 1×10^{-6} (one chance of health effect in one million) is generally considered by the U.S. EPA to be an acceptable level. Based on these risk values and the population distribution, the annual incidence of cancer due to dioxin and dibenzofuran ambient air exposure in Columbus is on the order of one case every 100 years (1 every 2000 years with proposed risk). The uncertainty of these risk calculations may easily be as great as one order of magnitude.

We conclude that given the current knowledge of the health effects of exposure to PCDD/PCDF in ambient air, there is no appreciable risk to public health from the concentrations of these compounds found in the ambient air in Ohio. However, a more important route of exposure may be from ingestion of contaminated milk and food products. Because these compounds have been found in human tissue, the importance of the ingestion route should be studied in greater detail.

ACKNOWLEDGEMENTS

This work was supported by the Ohio Air Quality Development Authority and the Ohio Environmental Protection Agency.

REFERENCES

1. C. Rappe and L.-O. Kjeller, "PCDDs and PCDFs in Environmental Samples: Air, Particulates, Sediments and Soil," **Chemosphere, 16:** 1775 (1987).

2. J. M. Czuczwa, B. D. McVeety, and R. A. Hites, "Polychlorinated Dibenzo-p-dioxins and Dibenzofurans in Sediments from Siskiwit Lake, Isle Royale, **Science, 226:** 568 (1984).

3. U.S. Environmental Protection Agency, "Health Assessment Document for Polychlorinated Dibenzo-p-Dioxins," EPA/600/8-84/014F, Environmental Criteria and Assessment Office, Cincinnati, Ohio (September, 1985).

4. R. R. Bumb, W. B. Crummett, S. S. Cutis, J. R. Gledhill, R. H. Hummel, R. O. Kagel, L. L. Lamparski, E. V. Luoma, D. L. Miller, T. J. Nestrick, L. A. Shadoff, R. H. Stehl, and J. S. Woods, "Trace Chemistries of Fire: A Source of Chlorinated Dioxins," **Science, 210:** 385, (1980).

5. U.S. Environmental Protection Agency, "Health Advisories for 25 Organics," PB87-235578, Office of Drinking Water, Washington, D.C. (March 1987).

6. J. M. Czuczwa and R. A. Hites, "Airborne Dioxins and Dibenzofurans: Sources and Fates," **Environ. Sci. Technol. 20:** 195 (1986).

7. B.D. Eitzer and R.A. Hites, "Dioxins and Furans in the Ambient Atmosphere: A Baseline Study," **Chemosphere,** in press (1988).

8. M. Buck and P. Kirschner, **Schrift. der Landesanstalt fur Immissions schutz des Landes NRW,** Heft 64: 164 (1986).

9. Christoffer Rappe and Lars-Owe Kjeller, "Identification and Quantification of PCDDs and PCDFs in Urban Air," **Chemosphere, 17:** 3 (1988).

10. K. Olie, V. D. Berg and O. Hutzinger, "Formation and Fate of PCDD and PCDF from Combustion Processes," **Chemosphere, 12:** 627 (1983).

11. B. J. Fairless, D. I. Bates, J. Hudson, R. D. Kleopfer, T. Holloway, D. A. Morey and T. Babb, "Procedures Used to Measure the Amount of 2.3.7.8-Tetrachlorodibenzo-p-Dioxin in the Ambient Air Near a Superfund Site Cleanup Operation," **Environ. Sci. Technol., 21:** 550 (1987).

12. U. S. Environmental Protection Agency, "National Dioxin Study Tier 4-Combustion sources, Project Summary Report," EPA-450/4-84-014g, Office of Air Quality and Planning, Research Triangle Park, NC (September, 1987).

13. P. C. Siebert, Denise R. Alston, J. F. Walsh and K. H. Jones, "Statistical Properties of Available Worldwide MSW Combustion Dioxin/Furan Emissions," presented at the 80th Annual APCA Meeting, Paper 87-94.1 (1987).

14. SIMCA-3B Pattern Recognition Programs; Principal Data Components: Columbia, MO.

15. Svante Wold and Michael Sjostrom, "SIMCA: A Method for Analyzing Chemical Data in Terms of Similarity and Analogy," in **Chemometrics: Theory and Application,** ACS Symposium Series 52: 283, American Chemical Society, Washington, D.C. (1977).

16. U.S. Environmental Protection Agency, "Municipal Waste Combustion Study: Emission Database for Municipal Waste Combustors," EPA/530-SW-87-021B, Office of Air and Radiation, Washington, D.C. (June, 1987).

17. R. E. Clement, H. M. Tosine, J. Osborne, V. Ozvacic, G. Wong and S. Thorndyke, "Emissions of Chlorinated Organics from a Municipal Sewage Sludge Burning Incinerator," **Chemosphere, 16:** 1985 (1987).

18. C. Rappe, L.O. Kjeller, P. Bruckmann and K.H. Hackhe, **Chemosphere,** in press.

19. J.G. Watson and T. Pace.   Chemical Mass Balance Model Version 6.0, provided by U.S. EPA and Desert Research Institute, Reno, NV (1987).

20. J.S. Bellin and D.G. Barnes, "Interim Procedures for Estimating Risks Associated with Exposures to Mixtures of Chlorinated Dibenzo-p-Dioxins and Dibenzofurans," EPA/625/3-87/012, March, 1987.

21. W. L. O'Connell, "A Review of the Emissions of PCDDs and PCDFs from Municipal Incinerators," Metcalf & Eddy, Inc., November, 1985.

22. U.S. Environmental Protection Agency.   "National Dioxin Study Tier 4 - Combustion Sources: Final Test Report - Site 1, Sewage sludge incinerator SSI-A," EPA-450/4-84-014J, April, 1987.

23. U.S. Environmental Protection Agency.   "National Dioxin Study Tier 4 - Combustion Sources: Final Test Report - Site 3, Sewage Sludge Incinerator SSI-B," EPA-450/4-84-0141, April, 1987.

24. D.J. Wackter and J.A. Foster, "Industrial Source Complex (ISC) Dispersion Model User's Guide - Second Edition," EPA-450/4-86-005a, June, 1986.

25. R.J. Kociba, D.G. Keyes, J.E. Beyer, R. M. Curreon, C. E. Wade, D. A. Dittenber, R. P. Kalnins, L. E. Frauson, C. N. Park, S. D. Barnard, R. A. Hummel and C. G. Humiston, "Results of a Two-Year Chronic Toxicity and Oncogenicity Study of 2,3,7,8-Tetrachlorodibenzo-p-Dioxin in Rats," **Toxicol. Appl. Pharmacol.**, **46(2):** 279, (1978).

26. R. M. Dowd, "EPA Revisits Dioxin Risks," **Environ. Sci. Technol.**, **22:** 373 (1988).

ASSESSING RISKS FROM INCINERATION OF MUNICIPAL SOLID WASTES

D.B. Chambers, B.G. Ibbotson and B.P. Powers

SENES Consultants Limited
52 West Beaver Creek Road
Unit #4
Richmond Hill, Ontario
L4B 1L9

ABSTRACT

Incineration likely will continue to play an important role in the management of municipal solid waste; however, there is much debate and concern over the potential risks posed by air emissions associated with the incineration of municipal solid waste. Using the "notorious" dioxin 2,3,7,8-TCDD and a reduced set of exposure pathways, this paper examines various issues concerning emissions from incinerators, notably the role of uncertainty and background exposures in the assessment of health risks. Various issues arising from recent proposals by the Enviromental Protection Agency for setting risk-based emission standards for hazardous air pollutants (NESHAPS) are also discussed.

INTRODUCTION

Although it is self-evident, we often forget that all human activities entail some risk. Kaplan and Garrick (1989) point out "...that we may make risk as small as we like by increasing the safeguards but may never, as a matter of principle, bring it to zero. Risk is never zero, but it can be small." While this statement is true it raises the question of what risks can be considered "small" or "safe" or "acceptable". Defining terms such as "safe" and "acceptable" has been debated extensively throughout recent hearings concerning vinyl chloride and subsequent EPA rulemakings under NESHAPS. On 20 July 1988, the EPA issued a proposed rule and notice of public hearing concerning the establishment of national emission standards for hazardous air pollutants (NESHAP). The notice proposed four policy approaches: Approach A (Case by Case based on all health information), Approach B (Incidence based on a level of 1 case per year), Approach C (1 x 10^{-4} Maximum Individual Risk), and Approach D (1 x 10^{-6}) Maximum Individual Risk).

In the notice the EPA noted, "Determinations on many of these issues within the proposed benzene regulation are expected to set precedents for the approach to be used for the substantial number of forthcoming NESHAP decisions" (53 FR 28502, 28 July 1988). EPA did in fact adopt these four approaches in its development of proposed rules for radionuclides under NESHAPS (54 FR 9612, March, 1989).

According to the July 1988 notice, the first step in the decision-making process is for the Administrator to determine a level of risk to health that is "safe" or "acceptable" (53 FR 28512). The Circuit Court in the decision on vinyl chloride does not specify what constitutes a "safe" or "acceptable" risk but does indicate that "safe" does not imply the elimination of all risk. The Federal Register notice, in referring to the Vinyl Chloride decision, notes (53 FR 28513):

> "There are many activities that we engage in every day - such as driving a car or even breathing city air - that entail some risk of accident or material health impairment; nevertheless, few people would consider those activities 'unsafe'."

Clearly, judgments are required as to the level of risk that can be considered "acceptable". The proposed rule notes that the determination of an "acceptable" risk requires consideration of the world in which we live and the health risks that our society faces.

An EPA memorandum to the docket entitled "Survey of Risks" (Docket No. QAQPS 79-3, Part 1, Docket Item X-B-1) summarizes various data on societal risks. One section of the memorandum sets out various "risk" benchmarks. For example, the risk associated with the EPA action level for radon is greater than 1×10^{-2}, the risk of dying in an automobile accident or by drowning is between 1×10^{-3} and 1×10^{-2}, and the FDA food tolerance level is set at 1×10^{-6}.

In addition to risk comparisons, decision makers can consider precedents established as part of other governmental regulatory actions. A recent paper by Travis *et al.* (1987), reviews 132 federal regulatory decisions in the area of cancer risk management and notes:

> "Two patterns are apparent. First, every chemical with an individual risk above 4×10^{-3} (four chances in 1000 that a chronically exposed individual will develop cancer) was regulated. Second, except for one FDA decision..., no action was taken to reduce individual lifetime risk levels that were below 1×10^{-6}."

Travis *et al.* note that EPA has suggested a *de minimis* individual lifetime risk level of 1×10^{-5} to 1×10^{-4} for small populations and 1×10^{-7} to 1×10^{-6} for large populations (ibid, p. 419). Travis *et al.* also indicate that for small population risks "the *de minimis* risk is now considered to be a 1×10^{-4} lifetime risk".

Other researchers have suggested that risk becomes insignificant at higher or lower levels, with many of these estimates lying in the range of 1×10^{-5} to 1×10^{-7} per year. All researchers agree that the appreciation of risk is highly subjective, varies from person to person, and depends on many factors. A *de minimis* value for the risk of death of 1×10^{-6} per year is one of the most frequently encountered values in the literature.

The concept of uncertainty is also integral to the meaning of the word "risk". Uncertainty arises in many ways, including uncertainty associated with the choice of mathematical model used in a simulation; uncertainty about the values of model parameters and uncertainty from the intrinsically stochastic behaviour of nature. For example, although we may know last year's rainfall, we are necessarily uncertain (due to natural variability) about the amount of rain that may fall next year. Risk analysis thus examines uncertain events with uncertain outcomes. Uncertainty about risks (that arises for example, in the extrapolation of animal data at high doses to man at low doses) further heightens the public concern over the potential risks.

In the final benzene rule issued in September 1989, EPA notes that:

> "EPA strives to provide maximum feasible protection against risks to health from hazardous air pollutants by (1) protecting the greatest number of persons possible to an individual lifetime risk no higher than approximately 1 in 1 million and (2) limiting to no higher than approximately 1 in 10 thousand the estimated risk that a person living near a plant would have if he or she were exposed to the maximum pollutant contaminant for 70 years" (ibid, p.5).

The EPA acknowledges that while the Agency can establish such goals as fixed numbers,

"...the state of the art of risk
assessment does not enable numerical risk
estimates to be made with comparable
confidence. Therefore, judgement must be
used in deciding how numerical risk estimates
are considered with respect to these goals"
(ibid, p.6).

The final benzene rule also acknowledges

"that consideration of maximum individual
risk ("MIR") - the estimated risk of
contracting cancer following a life time
exposure at the maximum modelled long-term
ambient concentration of a pollutant - must
take into account the strengths and
weaknesses of this measure of risk. It
is an estimate of the upper bound of risk
based on conservative assumptions, such as
continuous exposure for 24 hours per day for
70 years. As such, it does not necessarily
reflect the true risk, but displays a
conservative risk level which is an upper
bound that is unlikely to be exceeded."

In this regard, the Radiation Advisory Committee (RAC)
of the EPA Science Advisory Board which is commonly
referenced in the Background Information Document for
Prepared NESHAP's for Radionuclides (EPA, 1989) notes
various bias in the EPA exposure assessment (RAC, 1989, DP8-
10) and recommends that "Best estimates of doses and
risks with appropriate uncertainty statements should be
used in all risk assessments. The best estimates should
be statistically defined, according to the target population
or individual and the shape of the uncertainty
distributions" (RAC, 1989, p. 10). The RAC also
recommends that EPA should qualify the uncertainty in its
estimates of dose and risk (ibid, p. 13, 14).

We concur with this view and using 2,3,7,8-TCDD as the

basis for discussion, illustrate some aspects of such an analysis for air emissions from a municipal incinerator. The calculations were performed using a computer model developed at SENES. The Assessment of Incinerator Risk (AIR) model was designed to analyze long-term risks from both the direct and indirect exposure to local residents from incinerator air emissions. The AIR model has been used to assess several municipal and industrial incinerators and incorporates Monte Carlo methods to analyze uncertainties in exposure assessments.

EXISTING ENVIRONMENT

Individuals are exposed to a vast array of chemicals from both anthropogenic sources, such as exhaust from automobiles or industrial emissions, and non-anthropogenic sources, such as particulate matter and combustion products from forest fires. Travis and Hattemer-Frey (1989) have examined the literature on dioxin exposures due to emissions from municipal solid waste incinerators and natural sources. They conclude that background sources of dioxin account for more than 99% of the total daily intake of dioxin even to persons living near municipal incinerators (ibid, p. 95). While high levels of exposure to natural sources of dioxin do not justify additional high exposures due to human activity, consideration of natural exposures does however provide one perspective from which the total exposure (and risk) due to dioxin can be examined. Therefore, the present analysis in this paper estimates exposure and risks associated with dioxins emitted from a municipal solid waste incinerator with and without natural background exposures.

EXPOSURE PATHWAYS

There are numerous pathways by which people can be exposed to emissions from incinerator stacks. These include inhalation of dust or vapours, direct ingestion of

274

soil or household dust, the ingestion of plants grown locally in contaminated soil, the ingestion of drinking water from an adjacent water body, and the consumption of sport fish, meat and milk. In this paper, the reduced (i.e. partial) set of pathways shown in Figure 1 for an urban adult male are considered.

The AIR model provides for the analysis of risks due to inhalation and ingestion pathways over an exposure period of 70 years. It is assumed in this analysis that the receptor is an adult male between 20 and 39 years of age, who weighs 70 kg, has a breathing rate of 23 m^3 per day, and consumes 0.6 kg of produce per day.

The environmental fate of incinerator emissions and the exposure pathways considered in this analysis are briefly described below.

Air Concentration

Air dispersion analyses are performed with the ISC-LT model. This model has been validated by the EPA and versions of the model have been used in previous risk assessments of incinerators (e.g. Travis et al., 1987 and Radian, 1987). The ISC-LT model is a generalized Gaussian plume model that enables the user to calculate the impact of emissions on air quality from point and area sources. It is a sector-averaged model that builds upon the Air Quality Display Model (AQDM) and the Climatological Dispersion Model (CDM). Annual or seasonal statistical wind summaries (STAR data sets) are employed by ISC-LT to calculate air concentrations. A complete description of the capabilities and inputs required to run the model is available in the users manual (TRC, 1986).

Inhalation

Inhalation exposures occur continuously and result from the inhalation of total suspended particulate matter (TSP) and gases in the air. Exposures are estimated from the amount of air breathed and the calculated concentrations of TSP matter and gases both outdoors and indoors. The inhalation algorithms take account of different activity patterns in the winter and summer for both the adult and child receptors. The present analysis assumes that about 50% of the ambient TSP level is resuspended soil and that indoor TSP concentrations are about 75% of outdoor concentrations. For the present analysis, the

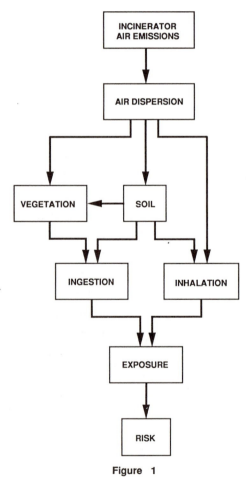

Figure 1

EXPOSURE PATHWAYS CONSIDERED

receptor is assumed to spend about 10% of his time out of doors in the summer time.

Soil Concentrations

The concentration of a chemical in soil (C_s) is estimated from assuming a steady state balance of atmospheric deposition, and removal via biodegradation and leaching of chemicals into the soil profile 1:

$$C_s = \frac{N_{total} \left[1 - \exp - k_{st}t\right]}{D_{soil} \ Z \ k_{st}} + C_{sb}$$

where:

C_s = concentration of chemical in soil (g/g)
N_{total} = ground-level deposition rate (g/m^2/d)
D_{soil} = soil bulk density (g/m^3)
Z = mixing depth (m)
k_{st} = total rate constant for environmental losses of the chemicals from soil (1/d)
t = time during which deposition occurs (d)
C_{sb} = background concentration in soil (g/g).

Ingestion of Dirt and Dust

Direct ingestion exposures stem from the inadvertent or deliberate ingestion of dirt and dust. While this pathway is particularly relevant to young children who may eat foreign material such as dirt and frequently put their fingers in their mouths, adults may also transfer dirt and dust from their hands to their mouths through activities such as eating or smoking, or after gardening for example.

In the present situation, the adult receptor is assumed to ingest the dirt from the inside surfaces of both hands, on summer days, when working outdoors. The inside surfaces of the adult's fingers and thumb account for about 14% of the area of hand (approximately 0.013 m^2 for both hands). In a similar manner, the adult is assumed to ingest indoor dust equivalent to the amount on one-half of the inside surfaces of his hands (Hawley, 1985). This results in an ingestion of approximately 110 mg of dust on those days spent indoors both summer and winter.

Vegetation (Produce)

Uptake by plants generally is considered to occur via three methods or modes: uptake through the root system, deposition of solid particles on leaves, and uptake of vapours through leaf pores. Uptake of vapours is not considered in this analysis.

Uptake through the root system is thought to be more important for highly soluble substances and for root crops such as carrots. Deposition of particulate matter on leaves is thought to be more important for less volatile chemicals and for leafy plants such as lettuce. The estimation of deposition is usually based on particle deposition velocities, duration of deposition, and processes which remove deposited matter from plant surfaces (such as rainfall).

The total uptake of a chemical by plants, i.e., root uptake and foliage deposition, is estimated using the simplified relation:

$$C_{PL} = C_a \, U_p + B_v \, C_{sl}$$

where:
C_{pl} = concentration of chemical in local plants (g/kg dry)
C_a = concentration of chemical in air ($\mu g/m^3$)
C_{sl} = concentration of chemical in soil (g/g)
B_v = transfer factor for soil to plants (g/kg dry)
U_p = deposition to plant (g/kg dry)

278

The receptor is assumed to live in a single-story house. Some of the fruits and vegetables he consumes, are assumed to come from an on-site garden.

The amounts and types of produce that people might consume is influenced by the size of the garden, the yields of the crops grown, and the preferences of the receptors. Information from surveys of household garden sizes, yield, and other factors such as food preferences are used when available. In the Toronto, Ontario area, typically about 5% of the produce consumed is home (or locally) grown. Travis and Hattemer-Fry (1989) also suggest that about 5% of vegetables produced and consumed in the U.S. are homegrown.

Exposure and Dose

Given the concentrations of chemicals in the various exposure media, the rates of inhalation and ingestion are calculated by estimating consumption patterns, and where appropriate, activity levels. Doses are calculated by adjusting exposure (via inhalation or ingestion) for pollutant bioavailability according to the following algorithm:

$$I = Q \times C$$

where:
I = amount of chemical ingested/inhaled (g/d)
Q = quantity of foodstuff ingested or
     air breathed (kg/d or m^3/d)
C = concentration of chemical
    ingested/inhaled (g/kg or g/m^3)

The total dose that a receptor receives is a function of the exposure, the bioavailability of the chemical, and the receptor's body weight:

$$D_t = \left(\frac{I_{ig} B_{ig} + I_{ih} B_{ih}}{B_w} \right)$$

where:

D_t = total individual dose (g/kg/d)
B_{ig} = bioavailability factor for ingestion (g/g)
B_{ih} = bioavailability factor for inhalation (g/g)
B_w = receptor's body weight (kg)

Risk

Dioxin (and furans) are not a single substance but a generic term used to describe a family of substances differing from one another by the number and position of chlorine atoms within the molecule. The dioxin isomer 2,3,7,8-TCDD is considered to be the most toxic. Table 1 shows the toxicity of the various dioxin congener groups relative to 2,3,7,8-TCDD (after NATO Report Number 176, 1988).

Table 1

International Toxicity Equivalency Factors (I-TEFs)
and Proportion of Congeners of Concern in a Homologous Group*

Congener of Concern	I-TEF	Congeners of Concern in a Homologous Group
2,3,7,8 TCDD	1	1 out of 22 (5%)
1,2,3,7,8 PeCDD	0.5	1 out of 14 (7%)
1,2,3,4,7,8 HxCDD 1,2,3,7,8,9 HxCDD 1,2,3,6,7,8 HxCDD	0.1	3 out of 10 (30%)
1,2,3,4,6,7,8 HpCDD	0.01	1 out of 2 (50%)
OCDD	0.001	1 out of 1 (100%)

* **Source** – *NATO Report 176, 1988.*

Data such as that collected through the Canadian National Incinerator Testing and Evaluation Program (NITEP, 1985 and 1986) indicate that not all dioxin congeners are released at the same rate and that 2,3,7,8-TCDD may only be a relatively small fraction of the total dioxin released. For example, according to NITEP (1985) results for normal incinerator operation, the dioxin homologue breakdown would be as follows:

Homologue	%
T4CDD	3
P5CDD	10
H6CDD	18
H6CDD	29
O8CDD	40
Total PCDD	100

Moreover, chloro-homologue studies (Eitzer and Hites, 1988) indicate that the chloro-homologue profile of dioxin changes as a dioxin mixture progresses from source to sink. Again, 2,3,7,8-TCDD is only a relatively small fraction of total dioxin. In view of this, it is clear from consideration of Table 1 that assuming all dioxin is 2,3,7,8-TCDD will overestimate the potential risk.

Human populations have been exposed to PCDDs (dioxins) and PCDFs (furans) for a number of years through such materials as chlorinated phenols, PCBs, phenoxy herbicides, incineration ash and various other chlorinated chemicals, yet no long-term adverse health effects have been positively correlated with these exposures (NRCC, 1981; MOE, 1985).

The understanding of the mechanism of action of PCDDs/PCDFs is important. The two-stage theory of carcinogenesis states that during the first stage there is an interaction of the chemical carcinogen with DNA. This interaction results in an altered cell and the process is known as initiation. The initiated cell can be exposed to agents which do not interact with DNA, but enhance the cellular response to the initiator and thereby cause an increase in tumour formation. Such agents are termed promoters, and chemicals that act as both promoters and initiators are known as complete carcinogens. Agents that increase tumour incidence without interaction with DNA may do so through suppression of the immune system, direct cytotoxicity, peroxisome proliferation, alterations in cellular membranes or other mechanisms of tumour promotion (Clayson, 1987; Shu *et al.*, 1987).

The Ontario Ministry of the Environment (and various European Governments) has concluded that PCDDs (and PCDFs) are non-genotoxic carcinogens and hence there is a threshold in the dose-response relationship. Applying a 100-fold uncertainty factor to a no-observed-adverse-effect-level (NOAEL) in rat bioassays, the Ontario Ministry of the Environment has calculated an acceptable daily intake (ADI) of 10 pg of 2,3,7,8-TCCD per kg body weight per day.

This runs counter to the views of the California Air Resources Board (CAPCD, 1987), EPA (1985 and 1987) and others who have concluded that there is no threshold dose and hence a risk of cancer at any level of exposure no matter how small. For this analysis, a risk factor of about 0.002 pg per kg body weight per day (derived from CAPCD, 1987) is assumed to result in a (lifetime) cancer risk of 1×10^{-6}.

Gough (1988) has reviewed the various approaches to estimating cancer risk associated with exposure to TCDD. He notes that until late 1987, all U.S. Regulatory agencies used no-threshold models to estimate cancer risk from dioxin. In 1987, EPA considered both threshold and non-threshold models and arrived at a value midway between the estimates generated by the two approaches. The EPA 1987 value of 0.1 pg per kg-body weight per day (for a risk of 1 x 10^{-6}) is 17 times higher than its previous estimate of 0.006 pg per kg-body weight per day (VSD) (Gough, 1988, p. 340). Gough finds the EPA revision unsatisfying as it is based on a meshing of two fundamentally different models. Finkel also argues against the averaging of results derived from "two fundamentally irreconcilable theories" (1988, p. 161).

Both types of risk prediction models are considered in the example calculations.

Uncertainty Analysis

Evaluating the potential risks of incinerator emissions requires mathematical models which simulate the environmental transport and exposure pathways of the various constituents. By their very nature, models are only partial representations of real systems and contain a number of parameters which are either imperfectly known or inherently variable by nature. Uncertainties also arise when these models are used to predict effects over conditions and periods of time different from those for which the models and model parameters were developed. Consequently, the results of all model calculations are uncertain to some degree. Monte Carlo methods can be used to assess uncertainty about the various estimates of concentrations, exposure and risk. Indeed, the California Air Pollution Control Districts (CAPCD, 1987) reference manual, indicates that such "probabilistic methods (e.g. Monte Carlo simulation) are state-of-the-art in evaluating uncertainty in a risk assessment".

RESULTS OF EXAMPLE CALCULATION

Objectives of Assessment

For this analysis, the maximum exposed individual (MEI) is assumed to live about one kilometre downwind in the prevailing wind direction from a large municipal incinerator. As indicated previously, the MEI is assumed to live at this location for his entire 70 year lifetime and remain at home 24 hours per day. He is outside about 10% of the time and works in a garden in which he grows a small portion of the fruit and vegetables he eats. At this point, the purpose of the risk assessment needs to be revisited. Most, but not all, of the factors noted above lead to conservative estimates of dose and risk which could be appropriate if the objective is to establish a conservative basis for a regulatory decision. On the other hand, if the objective of the assessment is to develop a best estimate of dose and risks that people are likely to experience, then several (but not all) of the foregoing assumptions are inappropriate.

If the objectives of the risk assessment are not clearly defined, a bias will creep into the exposure assessment. For example, consider the frequent assumption of a 70 year exposure duration. While lifetime occupancy in the same residence is possible, it is certainly not the norm. Recognizing the mobile nature of the modern work force, time away from home for advanced schooling or military service, changing residences in the same area (most people live in several residences over their lifetime), and even time spent at other locations during work, recreation or holidays, this assumption seems unrealistic. Data on human mobility indicate that there is only about a 0.04% chance that a person will remain at the same residence over his or her full lifetime (U.S. Department of Commerce, 1985). These same data indicate that people live an average of about 8.5 years at any one location. As there is a steep decline in concentration (and hence risk) with increasing distance from a source, unless the MEI moves to another house in the immediate neighbourhood, it is likely that changing residence would cause a reduction in exposure.

Thus, on average, the assumption of a lifetime (70

years) of exposure will overestimate lifetime risk (irrespective of the risk factors used in the analysis) by a factor of about 8 (i.e. 70 years/8.5 years).

Exposures by Pathway

The average intake and dose (i.e. intake times bioavailability factor) for each pathway of exposure considered in this analysis are given in Table 2. Where background levels are excluded from the calculation, the inhalation of dust and soil is predicted to be the largest exposure pathway, accounting for nearly 85% of the total dose. The ingestion of household dust and soil are the next two largest contributors to dose, each accounting for about 7% of the total. These results reflect the tendency of dioxin to become attached to airborne particles, to be deposited on soil, and to accumulate over time. The ratio of exposures due to ingestion of soil and household dust will vary somewhat depending on various factors used in the assessment such as the assumed reduction of indoor particulate levels over outdoor particulate levels and time spent indoors.

When background dioxin levels in soil and produce are included in the calculations the relative importance of the exposure pathways changes. Ingestion of dust and soil now each account for more than 40% of the total dose. Moreover, the average intake or exposure with background included is about 100 times greater than the intake or exposure calculated for the incinerator alone. If the amount of homegrown produce was assumed to be 50% (rather than 5%) of the produce consumed, the relative contribution of the produce pathway would increase to about 35% of the total, and if all of the produce consumed were assumed to be homegrown this pathway would increase to about 53% of the total.

Uncertainty Analysis

The parameters that were assumed to be uncertain in the

present analysis are shown in Table 3. Note that the
background dioxin concentration in air was assumed to be
zero for this example. Ontario data suggests that T4CDD
is present in urban air at levels below about 2×10^{-7}

Table 2

Summary of Exposure by Pathway for T4CDD*

Pathway		Average Intake $\mu g/day$	Average Dose $\mu g/kg/day$	Percent of Total Dose
Background Excluded	ingestion of dust	3.3×10^{-7}	7.0×10^{-10}	7.6
	ingestion of soil	3.0×10^{-7}	6.2×10^{-10}	6.8
	ingestion of produce	4.6×10^{-8}	9.6×10^{-11}	1.0
	inhalation of dust/soil	4.2×10^{-6}	7.7×10^{-9}	84.6
With Background and Soil and Produce	ingestion of dust	3.6×10^{-4}	7.4×10^{-7}	47.2
	ingestion of soil	3.2×10^{-4}	6.6×10^{-7}	42.1
	ingestion of produce	7.6×10^{-5}	1.5×10^{-7}	10.0
	inhalation of dust/soil	5.9×10^{-6}	1.1×10^{-8}	0.7

* Note — *Averages based on 200 random trials.*

$\mu g/m^3$. Thus at a breathing rate of 0.2 m^3/min, the MEI
would intake approximately 6×10^{-5} $\mu g/d$. This would be
about ten times the intake estimated for inhalation of
resuspended dust and soil, and about the same as that
attributed to produce. However, as T4CDD levels appear
to range from non-detectable values to a maximum of about

Table 3

Uncertain Parameters Used in Analysis

Module	Variable	[unit]	Distribution Type	Assumed Distribution *		
Air Dispersion	Emission value	$[g/s]$	Triangular	5.1×10^{-7}	5.1×10^{-6}	5.1×10^{-5}
Inhalation	TSP level	$[\mu g/m^3]$	Uniform	3.0×10^{-5}	8.0×10^{-5}	
Soil	Background concentration in soil	$[g/g]$	Uniform	0	9.0×10^{-9}	
Ingestion	See dose module					
Produce	Background produce concentration	$[g/g]$	Uniform	0	1×10^{-9}	
	Dry to wet conversion factor	$[g/g]$	Triangular	7.532	10.76	13.988
Dose	Bioavailability for ingestion	$[fraction]$	Triangular	0.01	0.125	0.30
	Bioavailability for inhalation	$[fraction]$	Triangular	0.01	0.075	0.30

* Note– Triangular distribution (minimum, mode, maximum). Uniform distribution (minimum, maximum).

2×10^{-7} µg/m^3 (CanTox, 1988; Environment Canada, 1988) the actual intakes from this source are likely to be smaller.

The exposure and risks predicted for the MEI for the pathways and parameters given above are summarized in Table 4. It is readily seen, that even with the limited number of parameters considered uncertain in this simplistic analysis that the minimum and maximum exposures and risks are different by about two orders of magnitude. As before, it can also be seen that exposures due to the incineration alone are only about 1% of those due to the incinerator and background (soil and produce) together.

For the very conservative risk factor used in this analysis (i.e. a risk of 1 x 10^{-6} for exposure to 2 x 10^{-9} µg per kg body weight per day) the best estimate of risk (here,

due to shape of distribution, the geometric mean) is about
4 x 10^{-6} without background and 6 x 10^{-4} with background
included. If the EPA 1987 VSD value of 0.1 x 10^{-6} µg per kg
body weight per day (for a risk of 1 x 10^{-6}) is used the mean
estimated risks would be smaller by a factor of 50, that is
about 8 x 10^{-8} without background and 12 x 10^{-6} with
background.

Table 4

Summary of Exposure and Risks to T4CDD[1]

	Parameter	Background Excluded	With Background
Total Exposure[2]	Arithmetic mean	9.1×10^{-9}	1.6×10^{-6}
	Standard deviation	7.3×10^{-9}	1.1×10^{-6}
	Geometric Mean	6.5×10^{-9}	1.1×10^{-6}
	Geometric standard deviation	2.4	2.6
	Minimum	5.5×10^{-10}	3.1×10^{-8}
	Maximum	3.5×10^{-8}	4.9×10^{-6}
Risk[3]	Arithmetic mean	5.1×10^{-6}	8.7×10^{-4}
	Standard deviation	4.1×10^{-6}	6.3×10^{-4}
	Geometric Mean	3.6×10^{-6}	6.2×10^{-4}
	Geometric standard deviation	2.4	2.6
	Minimum	3.1×10^{-7}	1.7×10^{-5}
	Maximum	1.9×10^{-5}	2.7×10^{-3}

Note – [1] *Statistics based on 200 random trials.*
[2] *Exposure in units of ug per kg body weight per day.*
[3] *Lifetime risk for lifetime exposure.*

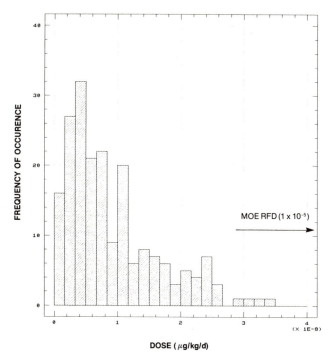

DOSE (μg/kg/d)

NOTE:

1.   REDUCED PATHWAYS
2.   BACKGROUND EXPOSURE EXCLUDED

Figure   2

**ESTIMATED INCREMENTAL DAILY EXPOSURE
TO T4CDD
EMITTED FROM INCINERATOR STACK**

The results of this limited uncertainty analysis are shown graphically in Figures 2, 3, 4 and 5. (The figures are histograms of 200 Monte Carlo trials.) Figures 2 and 3 show the predicted exposures without and with background, respectively. Figures 4 and 5 show the best descriptor. The graphs also demonstrate that uncertainty, even due to the relatively few parameters considered in this analysis, can affect the results and might well affect the assessment of whether or not emissions from a particular facility are or are not in compliance with a given criteria level.

The figures suggest lognormal behaviour and support the use of the geometric mean as perhaps the single predicted risks without and with background, respectively.

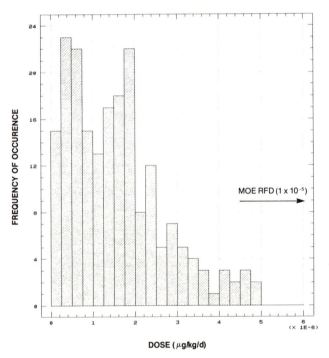

NOTE:

1.   REDUCED PATHWAYS

Figure 3

**ESTIMATED DAILY EXPOSURE TO T4CDD
DUE TO INCINERATOR AND BACKGROUND**

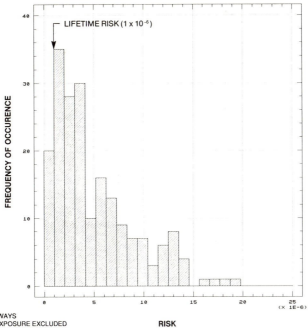

1. REDUCED PATHWAYS
2. BACKGROUND EXPOSURE EXCLUDED

RISK

Figure 4

ESTIMATED INCREMENTAL LIFETIME RISK
FROM EXPOSURE TO T4CDD
EMITTED FROM INCINERATOR STACK

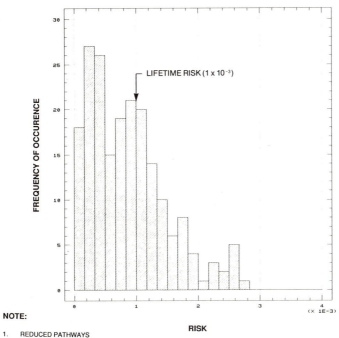

NOTE:

1. REDUCED PATHWAYS

RISK

Figure 5

ESTIMATED INCREMENTAL LIFETIME RISK
FROM EXPOSURE TO T4CDD
DUE TO INCINERATOR STACK AND BACKGROUND

The calculations also demonstrate the importance of second order (indirect) pathways and the significance of background sources of exposure.

Finally, the calculations illustrate the difference between the two approaches to assessing risk from TCDD. If TCDD is a threshold carcinogen (i.e. nongenotoxic) then people living in urban areas who are exposed to incinerator dioxin emissions would not be at risk of cancer according to these calculations as all exposures are below the threshold level. That is, the risk from TCDD would not only be small, it would for practical purposes be zero. On the other hand if TCDD is a nonthreshold carcinogen, then people would be subject to risks of the order of 10×10^{-6} (or greater) from natural sources of dioxins. The potential health and social implications of the two alternatives are clearly vastly different.

RECOMMENDATIONS

Albeit somewhat simplistic, the foregoing analysis has indicated that:

1. It is important to define the objective of a risk

    analysis and to select models and parameters consistent with those objectives.

2. Indirect exposure pathways have the potential to be as important or even more important than inhalation.

3. Incremental exposures (and risks) should be examined in the context of the world in which we live, that is in the context of natural exposures and risks. The present analysis suggests that incremental risks due to incinerator emissions of 2,3,7,8-TCDD are likely small compared to risks from exposure to natural sources of 2,3,7,8-TCDD.

4.   Uncertainty analysis should be performed routinely as part of risk assessment (as should sensitivity analysis to identify key factors) and a best or central estimate of risk should be presented along with a quantitative statement of uncertainty, (or uncertainty distribution) about that estimate.

5.   If it is not clear whether a substance should be assigned a threshold model or not, then both models should be evaluated independently.

MATERIALS CITED

California Air Pollution Control Districts (CAPCD), Interagency Working Group (1987), *Volume 1, Toxic Air Pollutant Source Assessment Manual for California Air Pollution Control District Permits.*

CanTox Inc., 1987. *Potential Health Hazard Assessment of Three City of Toronto Incinerator Facilities.* Prepared for the Department of Public Health, City of Toronto, October.

CanTox Inc., 1988. *Health Hazard Evaluation of Emissions from the Proposed PetroSun Inc./SNC Resource Recovery Incinerator.* Prepared for PetroSun Inc./SNC, February.

Clayson, D.B., 1987. *The Need for Biological Risk Assessment in Reaching Decisions about Carcinogens.* Mut Res 185:243-69.

Eitzer, B. and Hites, R.A., 1988. *Background Environmental Concentrations of Dioxins and Furans.* Proceedings of the 1988 EPA/APCA International Symposium: Measurement of Toxic and Related Air Pollutants. U.S. EPA/APCA UPIO (May), pp. 629-633.

Environ Corp., 1988. *Site Assessment, Phase 4B: Risk Assessment.* Prepared for the Ontario Waste Management Corporation (OWMC), January.

Environment Canada, 1985. *The National Incinerator Testing and Evaluation Program: Two-Stage Combustion (Price Edward Island)--Summary Report.* Urban Activities Division, Environmental Protection Service, September.

Environment Canada, 1988. *Detroit Incinerator Monitoring Program; Data Report No. 1: Windsor Air Sampling Site July 1987 - November 1987.* River Road Environmental Technology Centre, Ottawa, Ontario.

Finkel, A., 1988. *Dioxin: Are We Safer Now Than Before?* Risk Analysis, Vol. 8, No. 2, pp. 161-165.

Flakt Canada Ltd. and Environment Canada (Environment Canada), 1986. *The National Incinerator Testing and Evaluation Program: Air Pollution Control Technology - Summary Report.* September.

Gough, M., 1988. *Science Policy Choices and the Estimation of Cancer Risk Associated with Exposures to TCDD.* Risk Analysis, Vol. 8, No. 3, pp. 337-342.

Hawley, J.K., 1985. *Assessment of Health Risk from Exposure to Contaminated Soil.* Risk Analysis, Vol. 5, No. 4, pp. 289-302.

Holton, G., Little,C., O'Donnell, F., Etner, E., and Travis, C., 1981. *Initial Atmospheric Dispersion Modelling in Support of the Multiple-Site Incineration Study.* Oak Ridge National Lab., TN, ORNL/TM-8181.

Ministry of the Environment (MOE), 1986. *Scientific Criteria Document for Standard Development No 4-84, Polychlorinated Dibenzo-Dioxins (PCDD) and Polychlorinated Dibenzofurans (PCDF).* September.

NRCC, 1981. *Assessment of Toxicological Criteria. In: Polychlorinated Dibenzo-p-dioxins: Criteria for Their Effects on Man and His Environment.* Associate Committee on Scientific Criteria for Environmental Quality. Environmental Secretariat, National Research Council of Canada, Ottawa, Canada, p. 118.

Radian Corp., 1987. *Municipal Waste Combustion Study, Assessment of Health Risks Associated with Municipal Waste Combustion Emissions.* Prepared for U.S. Environmental Protection Agency, September.

Shu, H.P., Paustenbach, D.J., and Murray, F.J., 1987. *A Critical Evaluatin of the Use of Mutagenesis, Carcigenesis and Tumour Promotion Data in A Cancer Risk Assessment of 2,3,7,8-tetrachlorodibenzo-p-dioxin.* Reg Toxicol Pharm 7:57-88.

Travis, C.C., et al., 1987. *Potential Health Risk of Hazardous Waste Incineration.* Oak Ridge National Laboratories, TN, J. Hazard Mater. 14(3), pp. 309-320.

Travis, C.C., and Hattemer-Frey, H.A., 1987. *Human Exposure to 2,3,7,8-TCDD.* Chemosphere, Vol. 12: 2331-2342.

Travis, C.C. and Hattemer-Frey, H.A., 1989. *A Perspective on Dioxin Emissions from Municipal Solid Waste Incinerators.* Risk Analyses, Vol. 9 No. 1, pp. 91-97.

Travis, C.C., Richter, S.A., Crouch, E.A.C., Wilson, R. and Klema, E., 1987. *Cancer Risk Management, A Review of 132 Federal Regulatory Decisions.* Environmental Science and Technology, Vol. 21, no. 5, pp. 514-520.

CONNECTICUT'S DIOXIN AMBIENT AIR QUALITY STANDARD

Hari V. Rao and David R. Brown

Connecticut Department of Health Services
Division of Environmental Epidemiology and Occupational Health
150 Washington Street
Hartford, CT

ABSTRACT

Connecticut is the first state in the country to have adopted an ambient air quality standard for dioxins at 1 pg/m^3, 2,3,7,8-TCDD equivalents, as annual average. This paper describes the scientific basis and the methodology used by the State Department of Health Services (the risk assessment agency) in assisting the Department of Environmental Protection (the risk management agency) to establish a health-based dioxin standard. This standard protects the public health from the aggregate effect of all sources of dioxin emissions in the vapor and particulate phases. The risk assessment methodology included: a limit on total daily dioxin exposure from all media and sources based on reproductive effects, a multi-media non-source specific exposure assessment, an apportionment by media of the health-based limit (including background dosing rate), an evaluation of inhalation bioavailability and cancer risk based on a calculation of a range of upperbound cancer risk estimates using different potency, bioavailability, and particle phase assumptions.

INTRODUCTION

This report describes the scientific basis for Connecticut's primary Ambient Air Quality Standard (AAQS) for dioxins established by the State Department of Environmental Protection (DEP) at 1 picogram per cubic meter (1 pg/m^3) 2,3,7,8-TCDD equivalents as annual average (a picogram is one trillionth of a gram). Dioxins and 2,3,7,8-TCDD are used interchangeably in this report. The dioxin AAQS is based on the State Department of Health Services' (DHS) analysis of a Dioxin Health Risk Assessment prepared by Hart Associates.[1] The Health Risk Assessment was prepared in response to a legislative mandate, and designed to be consistent with DEP's Air Toxics Program requirements.[2] Thus, the standard setting process was a bilateral effort, which utilized the expertise of the DHS in risk assessment, and of the DEP in risk management. A Scientific Panel served as an advisory body.

Key Words. Reproductive Effects, Multi-media Exposure, Dose Apportionment, Bioavailability, Carcinogenic Potency.

The details of the Dioxin Risk Analysis, the key assumptions used, and their rationale are presented below:

DIOXIN DAILY DOSE AND BODY BURDEN

The risk analysis developed a rationale for consideration of daily dose and body burden for dioxins which are independent of specific sources. This was considered appropriate for the following reasons:

First, dioxin exposure is multimedia in nature. Dioxins have been detected in air, soil, sediments, suspended sediments, water, fish, meat, and milk as well as in human adipose tissue, breast milk, and blood.[3,4] The total body burden represents the sum of potentially significant individual contributions from the various media, sources and routes of human exposure. By comparing the relative contributions from each medium with the health-based limit, significant exposures can be identified and necessary measures taken to reduce such exposures.

Second, the evidence suggests that there exists a background body burden of dioxins in the general population of industrialized nations.[3] The body burden measurements provide an indication of past exposure, and can be used to calculate the total daily dioxin intake from the background exposure. The health impact from this daily background exposure can be assessed, and thus incorporated into the standard setting process. However, the ways in which all of the various media, sources and routes contribute to this background exposure remain unknown. Although there are uncertainties in assessing exposures, an advantage of this approach is that it places the various media of exposure, including the overall background multi-media exposure, in perspective.

REPRODUCTIVE EFFECTS AND DOSE LIMIT

Reproductive, developmental, and carcinogenic effects were determined to be health impacts of concern and were evaluated. Animal studies on 2,3,7,8,-TCDD toxicity have clearly demonstrated that it is a developmental and reproductive toxin in a variety of species at relatively low doses. A review of the pertinent developmental and reproductive studies can be found in EPA's Health Assessment Document for PCDDs.[5] The reported adverse outcomes include reduced fertility, litter size and survival, offspring body weight changes, as well as cleft palate and kidney abnormalities. Among the available studies, Murray et al's [6] 1979 study on Sprague Dawley rats, and Allen et al's [7] 1979 study on rhesus monkeys were considered appropriate for quantitative assessment.

Murray et al's study examined the effects of dietary exposure to 2,3,7,8-TCDD on reproduction in Sprague Dawley rats over three generations. The rats were given 0, 0.001, 0.01, and 0.1 ug/kg/day. No significant toxic effects were observed in the F_0 generation during 90 days treatment prior to mating. The study showed that the lowest dose, 0.001 ug/kg/day had no effect on fertility, litter size or fetal survival. The authors concluded that the doses 0.01 and 0.1 ug/kg/day produced significant effects on the reproductive capacity through three generations, F_0, F_1, and F_2. The study indicated that the 0.001 ug/kg/day could be considered as a no effect level. A reanalysis of the Murray et al data using a different statistical approach concluded the lowest dose was an effect level and that a no effect level could not be determined. [8] Since there was a question relative to the no effect level in the rat study, DHS considered the data on 2,3,7,8-TCDD's effects on reproduction in rhesus monkeys. The doses administered in the diet were 0, 1.8 (50 ppt) and 18 ng/kg/day (500 ppt) up to 9 months. Following 7 months of treatment, the females were mated with untreated males. At the higher dose (18 ng/kg/day), there was a decrease in serum estradiol and progesterone. The menstrual cycle was however not affected. Only three animals conceived, after which two aborted and one had a normal birth. At the 1.8 ng/kg/day dose serum estradiol and progesterone levels were normal. Eight treated females were mated with untreated males; there were six pregnancies, four abortions and two normal births. DHS judged that the results of the sensitive rhesus monkey studies could potentially support a more conservative effect level than that reported in the rat studies. Accordingly, the 1.8 ng/kg/day lowest effects level was used to calculate a health-based limit on the total dioxin daily intake from multimedia exposure.

Besides reproductive effects, tumorogenic effects have been observed in rodent experiments following chronic exposure to low levels of 2,3,7,8-TCDD. For example, Kociba et al's 1978 two year study tested cancer response in rats at doses of 1, 10, and 100 ng/kg/day. The 10 ng/kg/day dose caused a statistically significant increase in liver tumors in experimental animals versus controls. [9]

A comparison of the rodent dose-response data from the cancer study by Kociba [9] and reproductive study by Murray [6] showed that the two experimental adverse outcomes observed in separate bioassays, reduced fertility and liver tumors, appear to be the result of exposure to equi-toxic doses of 2,3,7,8-TCDD, i.e., 10 ng/kg/day. The experimental exposure durations were different, 3 months for first reproductive effects, and 24 months for tumor effects (time to fatal tumor data are not available). For a congener like 2,3,7,8-TCDD, the cumulative dose is more critical than the dose rate. [10] The data from reproductive and cancer bioassays support this view. By factoring the exposure duration and calculating the cumulative dose for each outcome, it can be shown that 12 to 25 percent of the cumulative cancer dose causes fertility effects in the same species. The cumulative dose analysis suggests that the potential adverse reproductive effects from dioxin exposure present a substantial immediate concern and cancer is a chronic concern at the same levels of exposure. The data from the rhesus monkey studies by Allen et al [7] provide further support to the argument that the reproductive response is a very sensitive response. The experimental evidence points to a lower Lowest Observed Effect Level (LOEL), 1.8 ng/kg/day, compared with a LOEL of 10 ng/kg/day, identified in the rat studies. Factoring these values, and the exposure duration of six months in the non-human primate studies, and three months in the rat reproductive bioassays, even a smaller fraction, approximately 40 percent, of the cumulative administered dose (rats) can be estimated to elicit adverse reproductive effects in rhesus monkeys, i.e., the latter species exhibits 2-3 fold greater sensitivity than the rodent species. The cumulative dose response analysis places the sequence of health concerns - reproductive, developmental and carcinogenic in perspective.

A further concern arose from the experimental observations of Moore et al which showed that high levels of the unmetabolized dioxin congener, 2,3,7,8-TCDD, were excreted in milk and that each rat pup actually received a higher dose during the first week after birth than was administered initially to the mother. [11] This study revealed that while TCDD crosses the placenta in the rat, exposure of the offspring occurs mainly through nursing. Thus the maternal milk pathway is a significant pathway affecting neonatal development in rats. Dioxins have been detected in human breast milk but there is no evidence to link dioxin exposure through nursing to human neonatal developmental toxicity. It was concluded that the animal data on reproductive effects can be used to derive a total dose limit to human exposure in 2,3,7,8 - TCDD equivalents. Thus, the reported LOEL 1.8 ng/kg/day in the rhesus monkey study was reduced by an Uncertainty Factor of 1000 to estimate the dose limit at 1.8 pg/kg/day.

DOSE APPORTIONMENT BY MEDIUM

Since cumulative dose is relevant to dioxin effects and since there are multiple media and sources of dioxins, it was necessary to consider an apportionment approach.

Assumption. Considering the multi-media nature of dioxin exposure, the health-based limit of 1.8 pg/kg/day (based on reproductive effects as most sensitive response) should be apportioned by medium of exposure and an allowable level established for each medium. This apportionment considers potential background exposures to be significant.

Rationale. Although the background dioxin levels in the environment contribute to the total human body burden, this information was not factored in the calculation of the health-based total dose limit of 1.8 pg/kg/day. Thus the limit reflects the total theoretical permissible daily dose of dioxins from all media of exposure, including background exposure. It represents the maximum daily dose that should not be exceeded to assure that no adverse health effects occur over a lifetime of exposure to dioxins. Therefore when assessing only one of the several possible exposure media, it is necessary to apportion the health-based limit to account for other potential exposures.

The first step in apportioning multi-media exposure of humans to dioxins was to estimate the background contribution to total dioxin exposure. The average daily intake of dioxin can be estimated using a linear, one compartment model: [3]

| Background Dose Rate | = | Body Burden x ln 2 / half-life |
| (ng/kg/day) | | (ng/kg)   (days) |

Assuming that a human weighs 60 kg, has 20 percent fat, and has 7 ppt dioxin in fat (ng/kg),[12] then the body burden of dioxin is 84.0 ng. The half life is assumed to be 5.8 years (2120 days)[13] and the dose rate is estimated to be approximately 0.45 pg/kg/day. Travis and Hattemer-Frey estimated through half-life modeling and 70 kg assumption that human exposure to 2,3,7,8-TCDD is about 0.4 pg/kg/day. [3] An EPA calculation showed a range of estimates of daily intake of 2,3,7,8 - TCDD between 0.04 to 0.51 pg/kg/day. The daily dioxin intake value used by DHS is in reasonable agreement with those reported above. This background exposure at 0.45 pg/kg/day (direct and indirect) represents 25 percent of the health-based limit of 1.8 pg/kg/day.

Estimates of the relative contributions from air, food, water, and soil to the daily human exposure to dioxins were calculated from literature data. The available data on dioxin exposures were reviewed, in particular the Federal Ontario dioxin exposure assessment document. [14] This document assumed that dioxins in the Ontario environment are principally from incineration processes. Based on concentrations and contact rate the relative contributions were estimated to be: air (60%), water (5%), soil (5%), and food (30%). DHS adjusted this apportionment to account for (i) potential beef and milk exposure, and (ii) sensitive sub-groups (infants and children - milk pathway). Thus DHS estimated the relative contribution to be: air (40%), water (5%), soil (5%), and food (50%) in the Connecticut environment.

The relative source contribution of 40 percent from the air medium was

derived from the worst case exposure assessment of the Ontario environment.[14] The Ontario data represent the measurements of stack air (there
was no ambient air data) and the levels found in samples of fish, pork,
poultry products, drinking water, human fat, and the soil in the vicinity of
an incinerator. The Canadian assessment used (i) an estimated maximum annual average ambient air concentration of 8.4 pg/m^3 TCDD equivalents (60%
apportioned intake); (ii) for water, a concentration of 0.002 ng/L TCDD
equivalents (5% intake); (iii) for soil, a level of 81.1 pg/g TCDD equivalents (5% intake); and (iv) for food consisting of fish, poultry, pork and
eggs, 29.6 pg/g (30% intake). No meat, milk and fruit analyses were provided in the Ontario analysis, consequently, the Connecticut food apportionment was adjusted to 50%, and air to 40%.

For the air medium (40% apportionment), the matrix was considered to
include both vapor and particulate phases (the ambient air quality standard
takes into account both phases). Dioxins and furans released from a variety
of combustion sources have been shown to exist in vapor and particulate
phases [15]. The vapor phase, as well as the particulate phase (assumed
to be 100 percent in the respirable range) represent an inhalation hazard.
Moreover, volatilization from the background and atmospheric transport of
these semivolatile organics can potentially add to inhalation exposure.
Based on sampling data and modeling, 2,3,7,8-TCDD in the urban air has been
reported to exist in the particulate phase between 40 and 80 percent [16]
whereas the octaisomer is 100 percent particle bound.

The vapor phase half-life through photolysis has been reported to be
under six hours, and for the particulate phase the half-life is several hun-
dred hours. At locations close to the spectrum of combustion sources exposure to vapor phase dioxins via inhalation can occur, in addition to direct
inhalation of the respirable particulate phase. The background levels of
dioxins in the vicinities of resource recovery facilities in Connecticut
have been measured. The values (48 hr average) are: Mean = 0.045 pg/m^3 $\pm$
0.77, Maximum = 0.719 pg/m^3, Range = 0.004 to 0.719 pg/m^3 dioxin equivalents, and N = 130. Fish samples (background monitoring) showed that the
levels ranged from 0.23 to 8.95 pg/g for TCDF, and from a method detection
limit of 0.05 to 6.15 pg/g for 2,3,7,8-TCDD. The monitored background data
for Connecticut, although limited, indicate that both the atmospheric and
food chain exposures are potentially significant human exposure pathways.

The settling velocity is an important factor in determining the signif-
cance of exposure pathways. Whereas the inhalation exposure pathway contributes to a constant absorbed dose (1 pg/m^3 x 20 m^3/day = 20 pg/day)
from the vapor and particulate phases (this inhaled dose is independent of
settling velocity), the dose estimate for the indirect food chain pathway is
dependent on the settling velocity assumption. For example the Hart analysis [1] showed that the food chain contribution increased with increasing
settling velocity from 56, 80, to 98 percent for settling velocities of
0.0003, 0.001, and 0.01 m/sec respectively, for the same ambient concentration.

The question arises as to the appropriate settling velocity to use in
dose calculation. Travis et al's 1987 analysis used a settling velocity of
0.0023 m/sec and 100 percent particle-phase distribution to estimate the
food pathway's contribution to total daily intake (98 percent). [3] On the
other hand, a settling velocity of 0.001 m/sec was used in the Hart document
to estimate the food chain contribution (80 percent). [1] Additionally, if
the particle phase distribution were to be factored into the calculation (40
to 80 percent for 2,3,7,8-TCDD) then the food chain pathway percent contribution would be in the 56 to 72 percent range. Connecticut's 40 percent relative source contribution from air, and 50 percent from food to the daily

dioxin intake are consistent with an average settling velocity of 0.001 m/sec, and 60/40 particle-vapor phase distribution assumptions. It should be emphasized that the exposure assessment and dose apportionment for the Connecticut assessment are based on the assumption that direct inhalation intake is from vapor and particulate phases, and the indirect intake is primarily from the particulate phase. The apportioned daily dosing rate associated with air exposure is 0.72 pg/kg/day (40 percent of 1.8 pg/kg/day). The equivalent dioxin concentration in ambient air is 2.2 pg/m^3 (0.72 pg/kg/day x 60 kg/20 m^3 per day). The calculations and conversions are based on the dose limit of 1.8 pg/kg/day.

DEP RISK MANAGEMENT

The DEP considered the DHS assessment, the Hart assessment and their factors in the risk management phase.

Connecticut DEP reviewed a range of health-based estimates for a dioxin equivalent AAQS - 0.1 to 2.2 pg/m^3. The lower bound value (0.1 pg/m^3) comes from the initial Hart analysis and the upper bound, from the DHS analysis. DEP decided to reduce by a factor of 2.2 the highest concentration in the range and derived a level of 1 pg/m^3. This dioxin equivalent concentration of 1 pg/m^3 was proposed and adopted as the AAQS.

The DEP management decision to apply an additional safety factor of 2.2 was based on the desire for an added margin of protection against potential carcinogenic and immunotoxic effects and on operating considerations. This safety factor assured that no exceedences of the health-based limit would occur through indirect and background exposures. According to DEP, the decision considered other management inputs, such as monitoring and enforcement as well as analytical and statistical considerations.

The following analysis explains the health rationale for the 2.2 factor and shows how the standard of 1 pg/m^3 is protective of human health: At the maximum ambient dioxin concentration of 1 pg/m^3 from all combustion sources, the daily inhaled dose can be estimated to be 0.33 pg/kg/day (60 kg human body weight and 20 m^3 air breathed in a day). This calculated dose represents about 18 percent of the health-based limit of 1.8 pg/kg/day. The Hart document provided an estimate of the indirect contribution from 1 pg/m^3 air dioxin concentration to be about 1.0 pg/kg/day (100 percent particle phase assumption and 0.001 m/sec settling velocity). This intake is 55 percent of the limit. Adjusting for 40 to 80 percent particle phase, the indirect dose can be estimated to be 0.4 to 0.8 pg/kg/day (22 to 44 percent of the limit). Additionally the background can potentially contribute to an estimated 0.45 pg/kg/day (25 percent of the limit). Thus, the direct (inhaled), indirect, and background intakes can contribute up to 65 to 98 percent of the health-based limit. If the 2.2 safety factor is not applied to the 2.2 pg/m^3 estimate, a potential doubling of the dose would occur and the target limit would be exceeded. A higher safety factor (10) was considered not necessary since DEP proposed to regulate individual sources through an emission standard that ensures each source will have an insignificant impact on ambient air. The dioxin AAQS 1 pg/m^3 is thus considered to be protective of public health.

DIOXIN EXPOSURE AND CANCER RISK

The potential cancer risks from chronic exposure to dioxins via ingestion (indirect pathway) and inhalation (direct pathway) were evaluated. The calculation used the standard approach that the product of the exposure dose (pg/kg/day) and the potency value represents the potential upperbound risk.

Ingestion Risk. The ingestion risks from the indirect exposure pathway were estimated for three deposition scenarios. The calculation assumed that 100, 80, or 40 percent of the dioxin in the ambient air is in the particle phase and that the settling velocity is 0.001 m/sec. The estimated doses, 1.0, 0.8, and 0.4 pg/kg/day were multiplied by the oral potency values to calculate the potential upperbound risks, as shown in Table 1.

Inhalation Risks. Inhalation risk assessment for the direct exposure pathway focussed on the administered dose via inhalation, 2,3,7,8,-TCDD's potency via inhalation, and the effect of the matrix factor on dioxin's potency.

Administered Dose. The concentration in the ambient air (pg/m^3) and the contact rate (20 m^3/day) for a human body weight assumption (60 kg), and 100 percent pulmonary absorption determined the administered dose (pg/kg/day) via inhalation. A second calculation considered the particle/vapor nature of the airborne dioxins as well as the matrix effect and used an absorption factor of 50 percent to calculate the administered dose.

Inhalation Potency and Matrix Effect. Since 2,3,7,8-TCDD's potencies for the vapor and particulate phases via inhalation are not known, the oral potency values are used to predict the inhalation risk. This is considered a conservative approach for the following reasons:

Dioxin's potency estimates are based on the administered dose. In the lifetime feeding study the rats were given a diet mixed with dioxin dissolved in acetone matrix. [9] The bioavailability of dioxin in this matrix is reported to be 85 percent, compared with the values, 25 to 50 percent for the soil matrix, [17, 18] and 1 to 4 percent for the flyash matrix. [19] These bioavailability values are for oral/gastrointestinal absorption. The absorption values for the soil and flyash matrix indicate that considerably higher doses, 2 to 20 times higher than the dose in acetone matrix are required to produce the same dioxin concentration in the liver and elicit a tumor response. Such a shift in the dose/response curve relative to the matrix effect would predict a lower potency estimate for ingested dioxin. Thus the use of the oral potency values (acetone matrix) to predict the inhalation risk, particularly for inhaling particle bound dioxins, is considered conservative. Table 1 presents the estimated inhalation risks for 100 and 50 percent absorption assumptions.

Table 1. Dioxin Exposure and Upperbound Cancer Risk Estimates

| Dose * (pg/kg/day) | Assumption | 10^{-6} Risk Specific Dose (fg/kg/day) ** | | | |
		EPA 6	CT 36	FDA 60	EPA *** 100
Direct					
0.33	100% Abs	5.5×10^{-5}	9×10^{-6}	5.5×10^{-6}	3.3×10^{-6}
0.17	50% Abs	2.8×10^{-5}	4.5×10^{-6}	2.8×10^{-6}	1.7×10^{-6}
Indirect					
1.00	100% part	1.7×10^{-4}	2.8×10^{-5}	1.6×10^{-5}	1.0×10^{-5}
0.80	80% part	1.3×10^{-4}	2.2×10^{-5}	1.3×10^{-5}	8×10^{-6}
0.40	40% part	7×10^{-5}	1.1×10^{-5}	7×10^{-6}	4×10^{-6}

*   The dose was estimated for a dioxin concentration of 1 pg/m^3, a
     breathing rate of 20 m^3/day and a human body weight of 60 kg.
**   Femtogram (one quadrillionth of a gram).
*** Proposed potency estimate.

Depending on the scientific assumptions used and the estimate of the carcinogenic potency of dioxin, the range of upperbounds on total risk from indirect and direct exposure to 1 pg/m^3 ranges from 2 x 10^{-4} (5.5 x 10^{-5} + 1.7 x 10^{-4}) to 6 x 10^{-6} (1.7 x 10^{-6} + 4 x 10^{-6}). In the scientific judgement of DHS, the risk was estimated to be at the lower end of the range or 6 x 10^{-6}. A potential risk in the 10^{-6} range is considered acceptable for an aggregate air standard for all sources of dioxin emissions. However, for individual Resource Recovery Facilities, the potential cancer risk from ambient impact is even lower and is in the 10^{-7} range. The background estimate (0.45 pg/kg/day) represents a risk in the 10^{-5} - 10^{-6} range, depending upon the assumptions used.

DISCUSSION AND CONCLUSION

Connecticut is the first state in the country to adopt a dioxin AAQS that protects the public health from the combined effects of all sources of dioxin emissions. The AAQS is an aggregate standard and is different from the "standards" of other states which are in fact only maximum allowable impacts for individual RRFs. According to Connecticut DEP, "no resource recovery facility in the state is predicted to have an ambient impact of more than 0.037 pg/m^3 dioxin equivalents." [2] This maximum predicted impact for each RRF represents about 4 percent of the AAQS, and is consistent with the limits imposed by other states (Massachusetts 0.15, Pennsylvania 0.3, Rhode Island 0.02 to 0.2, and New Hampshire 0.09 to 0.27 pg/m^3). The predicted maximum impact for each RRF in the state represents a potential upperbound carcinogenic risk in the 10^{-6} to 10^{-7} range. No adverse reproductive and immunological effects are expected to occur at this impact level. The calculated dose for 0.037 pg/m^3 impact level (direct and indirect) is 0.049 pg/kg/day and it represents about 3 percent of the health-based limit, compared with the background's 25 percent. Clearly, there is a need to identify the sources contributing to the considerable background intake, and minimize such exposure.

The non-source specific approach used in the risk assessment is a departure from incinerator-specific risk assessments. Appropriately, this approach takes into account the body burden and daily dose from all media and sources as well as the background exposures when estimating a total daily dose and comparing with the target dose limit. The target dose limit assumption facilitated apportioning the dose by media. While developing a rationale for the 40 percent air apportionment, it became apparent that the risk assessment is sensitive to the settling velocity assumption used. Inhalation (44 percent) and meat/milk ingestion (56 percent) are significant exposure pathways for a minimum particle settling velocity (0.0003 m/sec) assumption. [1] The relative significance of the inhalation exposure decreases as the settling velocity increases. At the settling velocity 0.01 m/sec, the indirect food chain pathway dominates (98 percent) although the concentration of dioxins in the air is the same. Therefore, risk assessments of this type should estimate exposures based upon several settling velocities as a means of estimating a range of potential exposures. This is particularly necessary since the location of the plant affects the settling velocity. This type of analysis will help the risk manager to assess inhalation exposure separate from indirect multiple pathway exposure, if desired. A further improvement in dioxin risk assessment would come from the knowledge of vapor/particulate phase distribution of dioxins in ambient air. This information would improve the analysis of the indirect exposure pathways, since the deposition of dioxins from the ambient air has been shown to be particle phase dominated. [16]

REFERENCES

1.   F. C. Hart Associates, "Multiple Pathway Human Exposure and Health Risk
      Assessment of Polychlorinated Dibenzo-p-Dioxins and Polychlorinated
      Dibenzofurans from Municipal Solid Waste Incinerators," Prepared for
      State of Connecticut, Department of Health Services, February, 1987.

2.   Connecticut Department of Environmental Protection (DEP), "Basis for
      Standards and Procedures and Response to Comments on Proposed
      Resource Recovery Regulations (Air Pollution Provisions)," (1988).

3.   C. C. Travis and H. A. Hattemer-Frey, "Human Exposure to 2,3,7,8-TCDD,"
      Chemosphere 16, 2331-2342 (1987).

4.   C. C. Travis and H. A. Hattemer-Frey, "A Perspective on Dioxin
      Emissions from Municipal Solid Waste Incinerators," Risk Analysis 9,
      91-97 (1989).

5.   Environmental Protection Agency, "Health Assessment Document for
      Polychlorinated Dibenzo-p-dioxins," Office of Health and
      Environmental Assessment, May 1984.

6.   F. J. Murray, F. A. Smith, K. D. Nitschke, C. G. Humiston, R. J.
      Kociba, and B. A. Schwetz, "Three-Generation Reproduction Study of
      Rats given 2,3,7,8-Tetrachlorodibenzo-p-dioxin in the Diet,"
      Toxicology and Applied Pharmacology 50, 241-252 (1979).

7.   J. R. Allen, D. A. Barsotti, L. K. Lambrecht, and J. P. Van Miller,
      "Reproductive Effects of Halogenated Aromatic Hydrocarbons on
      Nonhuman Primates," Annals of New York Academy of Sciences 320,
      419-425 (1979).

8.   I. C. T. Nisbet and M. B. Paxton, "Statistical Aspects of three
      generation studies of the reproductive toxicity of 2,3,7,8-TCDD and
      2,4,5-T," The American Statistician 36, 290-298 (1982).

9.   R. J. Kociba, D. G. Keyes, J. E. Beyer, R. M. Carreon, E. E. Wade,
      D. A. Dittenber, R. P. Kalnins, L. E. Frauson, C. N. Park, S. D.
      Barnard, R. A. Hummel, and C. G. Humiston, "Results of a Two-Year
      Chronic Toxicity and Oncogenicity Study of 2,3,7,8-TCDD," Toxicology
      and Applied Pharmacology 46, 279-303 (1978).

10.   T. H. Umbreit, E. J. Hesse, and M. S. Gallo, "Reproductive Toxicity in
      Female Mice of Dioxin-Contaminated Soils from a 2,4,5 -
      Trichlorophenoxyacetic Acid Manufacturing Site," Archives of
      Environmental Contamination and Toxicology 16, 461-466 (1987).

11.   J. A. Moore, M. W. Harris, and P. W. Albro, "Tissue Distribution of
      14C Tetrachlorodibenzo-p-dioxin in Pregnant and Neonatal Rats,"
      Toxicology and Applied Pharmacology 37, 146-147 (1976).

12.   D. G. Patterson, J. S. Holler, S. J. Smith, J. A. Liddle, E. J. Sampson
      and L. L. Needham, "Human Adipose Data for 2,3,7,8-TCDD in Certain
      U.S. Samples," Chemosphere 15, 2055-2060 (1986).

13.   H. Poiger and C. Schlatter, "Pharmacokinetics of 2,3,7,8 - TCDD in
      Man," Chemosphere 15, 1489-1494 (1986).

14. Ontario Ministry of The Environment, "Scientific Criteria Document for Standard Development No. 4-84 – Polychlorinated Dibenzo-p-Dioxins and Polychlorinated Dibenzofurans," Ministry of The Environment, Toronto, March, 1986.

15. T. F. Bidleman, "Atmospheric Processes," Environ. Sci. Technol 22, 361–367 (1988).

16. T. F. Bidleman, "Gas-Particle Distribution and Atmospheric Deposition of Semi Volatile Organic Compounds," Presented at the EPA/ORNL Workshop on Risk Assessment for Municipal Waste Combustion, June 8-9, 1989.

17. T. H. Umbreit, E. J. Hesse, and M. S. Gallo, "Bioavailability of Dioxin in Soil from a 2,4,5-T Manufacturing Site," Science 232, 497–499 (1986a).

18. T. H. Umbreit, E. J. Hesse, and M. S. Gallo, "Comparative Toxicity of TCDD Contaminated Soils from Times Beach, Missouri, and Newark, New Jersey," Chemosphere 15, 2121–2124 (1986b).

19. M. Van den Berg, K. Olie, and O. Hutzinger, "Uptake and Selective Retention in Rats of Orally Administered Chlorinated Dioxins and Dibenzofurans from Fly Ash and Fly Ash Extract," Chemosphere 12, 537–544 (1983).

CONTRIBUTORS

Dr. Eros Bacci
University di Siena
Via delle Cerchia 3
53100 Siena, Italy
577-298836

Mr. Greg Belcher
Martin Marietta Energy Systems, Inc.
Y-12 Plant
P. O. Box 2009, MS-8227, Maxima Bldg.
Oak Ridge, TN 37831-8227
(615) 574-8203 or FTS 624-8203

Dr. Terry Bidleman
Chemistry Department
University of South Carolina
Columbia, SC 29208
(803) 777-4239

Dr. Roger Brower
Versar, Inc.
9200 Rumsey Road
Columbia, MD 21045
(301) 964-9200

Dr. David Brown
Environmental Epidemiology
 & Occupational Health
Connecticut Dept. of Health Services
150 Washington Street
Hartford, CT 06106
(203) 566-8167

Mr. Randy Bruins
US EPA, Region 10
MC-ES096
1200 Sixth Avenue
Seattle, WA 98101
(206) 442-2146 or FTS 399-2146

Dr. Davide Calamari
University di Siena
Via delle Cerchia 3
53100 Siena, Italy
577-298836

Dr. Sally Campbell
S.A. Campbell Associates
10714 Midsummer Lane
Columbia, MD 21044
(301) 730-6251

Dr. Douglas Chambers, Vice President
SENES Consultants Limited
52 West Beaver Creek Road, Unit #4
Richmond Hill, Ontario
Canada L4B 1L9
(416) 764-9380

Dr. Jean Czuczwa
Senior Research Chemist
Babcock & Wilcox
1562 Beeson Street
Alliance, OH 44601
(216) 829-7736

Dr. Sylvia Edgerton
Edgerton Environmental
412 Iliaina Street
Kailua, Hawaii 96734
(808) 254-6629

Ms. Holly Hattemer-Frey
Lee Wan & Associates
120 S. Jefferson Circle, Suite 100
Oak Ridge, TN 37830
(615) 483-9870

Dr. Robert Hodanbosi
Division of Air Pollution and Control
Ohio EPA
1800 Watermark Drive, Box 1049
Columbus, OH 43266-0149
(614) 644-3020

Mr. Brett Ibbotson, Principal
Angus Environmental Limited
1127 Leslie Street
Don Mills, Ontario Canada M3C 2J6
(416) 443-8362

Mr. Adam Kahn
Environ Corporation
210 Carnegie Center, Suite 201
Princeton, NJ 08540
(609) 452-9000

Mr. Ray Kapahi
EMCOM Associates
1433 North Market Blvd.
Sacramento, CA 95834
(916) 928-3300

Dr. Paul Koval
Air Pollution Toxicologist
Ohio EPA
Division of Air Pollution Control
P. O. Box 1049
Columbus, OH 43266-0149
(614) 644-2270

Dr. Donald Mackay
Department of Chemical Engineering
University of Toronto
Toronto, Ontario, Canada M5S1A4
(416) 978-5873

Dr. Craig Mc Farlane
US EPA
OEPER-Corvallis Environmental Research Lab
200 Southwest 35th Street
Corvallis, OR 97333
(505) 757-4670

Dr. Thomas McKone
University of California
Lawrence Livermore National Laboratory
P. O. Box 5507 (L-453)
Livermore, CA 94550
(415) 422-7535

Mr. Michael Marchlik
Ebasco Services, Inc.
Two World Trade Center
New York, NY 10048
(212) 839-3242

Mr. Obed Odoemelam
California Energy Commission
1516 9th Street
Sacramento, CA 95814
(916) 324-3590

Mrs. Sally Paterson
Department of Chemical Engineering
University of Toronto
Toronto, Ontario, Canada M5S1A4
(416) 978-5873

Mr. Bruce Powers
Senior Environmental Scientist
Angus Environmental Limited
1127 Leslie Street
Don Mills, Ontario Canada M3C 2J6
(416) 443-8362

Dr. Hari Rao
Environmental Epidemiology &
 Occupational Health
Connecticut Dept. of Health Services
150 Washington Street
Hartford, CT 06106
(203) 566-8167

Dr. Jerry Rench
SRA Technologies
4700 King Street
Suite 300
Alexandria, VA 22302
(703) 671-7171

Mr. Paul Shulec
Consulting Toxicologist
3600 Data Drive
#105
Rancho Cordova, CA 95670
(916) 852-0209

Dr. James Teitt
ERM, Inc.
116 Defense Highway, Suite 300
Annapolis, MD 21401
(301) 266-0006

Dr. Curtis C. Travis, Head
Risk Analysis Section
Health and Safety Research Division
Oak Ridge National Laboratory
P. O. Box 2008, MS-6109, 4500S
Oak Ridge, TN 37831-6109
(615) 576-2107 or FTS 626-2107

Mr. Rick Tyler
California Energy Commission
1516 9th Street
Sacramento, CA 95814
(916) 324-3590

Mr. Steve Washburn
Environ Corporation
210 Carnegie Center
Suite 201
Princeton, NJ 08540
(609) 452-9000

Mr. Mark Yambert
C&TD
Oak Ridge National Laboratory
P. O. Box 2008, MS-6243, 4500N
Oak Ridge, TN 37831-6243
(615) 574-0276 or FTS 624-0276

Dr. Kenneth Zankel, Manager
Permitting & Compliance, Atmospheric
 Sciences
Versar, Inc.
9200 Rumsey Road
Columbia, MD 21045
(301) 964-9200

INDEX